Selling Daylight

Praise for this book

'This is a wonderful book. Based on his experiences of nearly twenty years working in the renewable energy sector, Martin Bellamy offers a compelling vision of how what he perceives as a criminal under-utilization of PV across the developing world could be more forcefully and comprehensively addressed. You might not agree with everything he has to say but this is a must read for anyone interested in our energy futures.'

Ed Brown, Professor of Global Energy Challenges at Loughborough University and Co-Coordinator of the UK Low Carbon Energy for Development Network.

'Martin and this book are a gift to the world's unelectrified. Herein lie many key insights and a guide towards universal access to energy that changes lives across the globe. Read it – you won't be disappointed.'

Stephen Katsaros, CEO, Nokero Solar

Selling Daylight
A commercial strategy to address global energy poverty

Second Edition

Martin Bellamy

Practical Action Publishing Ltd
27a Albert Street, Rugby, CV21 2SG, UK
www.practicalactionpublishing.org

First published by Blurb Books 2017
Second edition published by Practical Action Publishing 2019

A catalogue record for this book is available from the British Library.
A catalogue record for this book has been requested from the Library of Congress.

978-1-78853-067-5 Hardback
978-1-78853-068-2 Paperback
978-1-78853-069-9 Epub
978-1-78853-070-5 PDF

Citation: Bellamy, M. (2019) *Selling Daylight: A commercial strategy to address global energy poverty*, 2nd edition, Rugby, UK: Practical Action Publishing <http://dx.doi.org/10.3362/9781788530705>

Since 1974, Practical Action Publishing has published and disseminated books and information in support of international development work throughout the world. Practical Action Publishing is a trading name of Practical Action Publishing Ltd (Company Reg. No. 1159018), the wholly owned publishing company of Practical Action. Practical Action Publishing trades only in support of its parent charity objectives and any profits are covenanted back to Practical Action (Charity Reg. No. 247257, Group VAT Registration No. 880 9924 76).

Copy-edited by Richard Barrett
Cover design by Danny Kwong http://dannykwong.com

Printed in the United Kingdom

Contents

About the author   x
Acknowledgements   x
Foreword   xi
Summary   xiii
The vision as narrative   xiii
Structure   xiii
Introduction   xvii

Part One   How we use energy and why change is paramount   1

1   **Electricity and people**   3
    Utility grid electricity   4
    On the grid or off the grid: four electricity areas   5
    Energy poverty   7

2   **Fuel and our lack of awareness**   11
    Our disconnection from fuel   11
    Central generation   12
    The burden of fuel   15
    What is electricity anyway?   16
    Changing our motivations   19

3   **The developing world cannot have reliable grid electricity** 21
    Keeping the lights on   21
    Realities of the grid   24
    Harming the developing world   27
    What societies demand   28

4   **Accepting that the grid format cannot end energy poverty**   31
    The 'alternatives' dilemma   31
    Knowledge and the policy makers   34
    The challenges of distribution   35
    Thinking about being proactive?   37
    Learning from the grid   39

Part Two   Introducing the Selling Daylight proposition   45

5   **Personalising energy**   47
    Involving the user in generation and consumption   48
    Promoting the value of energy services   50

http://dx.doi.org/10.3362/9781788530705.000

Sharing knowledge   53
Personalising hardware and knowledge   54

6   An introduction to PV systems   57
Grid-connected PV systems   58
'Off-grid' PV systems   61
A summary of PV systems   66
Generic services and sectors of society   67
Off-grid PV and personalisation   69

7   Universally accessible visual media   71
Why do we need visual media?   72
Some challenges of visual media   72
Foundation knowledge and value   77
The reach of visual media   79

8   The core value and poor reputation of 'off-grid' PV   81
The origins of off-grid PV's reputation   82
Bespoke systems   83
Mini-grids   85
Comparing industrial and public applications   87
Autonomous PV applications in industrialised societies   90
The legacy of PV technology   92

Part Three   Major elements of the Selling Daylight strategy   93

9   Focusing on quality and value   95
Developing nations and off-grid PV   95
Economic value   99
In search of quality   100
Elements of quality and value   102

10 Optimising stand-alone PV hardware design   105
Holistic solutions   106
The value of energy-efficient equipment   106
Integrated design   108
Involving the user   109
The enormous value of recorded data   113

11 Sharing essential knowledge with everyone   117
People we aim to reach   118
Training and investment   122
Public confidence   123

12 The need to reform procurement of stand-alone PV projects   125
Talking to the right people   126
Procurement specifications   126
Dynamic modification: readiness for change   130

A model of procurement   132
Procurement and value   133

13 Why creating an energy services brand is essential   135
The breakthrough brand?   137
Riches off the grid: why we need brands   138
Branding and value   139
The service and the brand   140

Part Four   People are the cause of poverty; and the solution   143

14 Why poverty persists   145
People: a perspective on poverty   147
Quantifying poverty   149
Why poverty persists   151
Affluence in a poor land   154

15 Finance, technology and people   159
Access to finance   159
Energy poverty and people   163
The attributes of energy poverty   154

16 Scale, diversity and people   169
The numbers describing poverty   169
Making sense of large numbers   171
The meaning of diversity   172
The people problem   174
Motivating millions of people   177

17 Motivating millions of people and giant companies   179
'Selfish desire' can ensure sustained effort   179
Scale, diversity and 'Personalised Energy'   181
The low-cost myth   188
How to sell quality autonomous energy services   191

Part Five   Implementing the Selling Daylight strategy   193

18 The need for a long-term strategy   195
Making the commitment   195
Feasibility and confidence   198
Investors waiting for certainty that will not come   200
The need to diversify   202

19 Why and how to be a diverse company   205
PV as a consumer product format   205
Physical versatility   209
Bridging global markets   213
Commercial momentum   214

20 Mapping global markets   217
    The industrialised world   218
    Disaster and emergency response   219
    Small Island States   220
    The developing world market   222
    Cumulative confidence, cumulative success   226

21 Where to start   229
    The markets, the risks, the lack of statistics   229
    The grid as a starting point   230
    Unexpected customers   233
    Categorising the commercial landscape   234
    Wants and needs   236
    Multinationals   238

Part Six   The commercial value and need for local partners   241

22 Selling guaranteed energy services   243
    Manufacturers and local people   244
    Reliability and value   246
    The incentive of product performance   250

23 Commercial growth from listening to the local market   255
    Market-driven product development   255
    Understanding local perceptions   259
    A 'solar light' and a 'solar home system'   260
    An assessment of spending   262

24 The essential role of local partner networks   265
    Increasing profit through partnership   265
    Existing local solar companies   267
    The realities of local logistics   271
    People support   273
    Challenges and opportunities   273

25 Global organisations partnering local networks   277
    The elements of growth   277
    Capital funding from commercial organisations   280
    The example of television   282
    Overcoming capital cost   284
    Financially retaining local partners   285
    The goal of global local networks   286

Part Seven   Why this strategy can work   289

26 A critical look at the Selling Daylight proposition   291
    A reality check   292
    Elephants in the room   293

Level heads and ego trips   295
Alternatives and the environment   297
Long-termism   299

27 A focus on value justifies the price   301
How do we value autonomous energy services?   301
ICT and healthcare   303
Urbanisation and its discontents   305
Selling movement   307
A force for democracy   311
Absolutely enormous potential   312

28 A civil dawn: life with inclusive autonomous energy services   315

References   321
Index   323

About the author

Martin Bellamy has 30 years of practical engineering experience. He joined the UK renewable energy industry in early 1999 and has focused on the design and provision of stand-alone ('off-grid') solar photovoltaic (PV) systems ever since.

His work spans the whole range of autonomous PV applications, from explosion-proof systems for industry to domestic services for energy-poor regions.

Martin specialises in developing world energy solutions. He provides product assessment, market insight and strategic business services to stand-alone energy organisations and the international investment sector.

He holds an honours degree in engineering physics and is chartered as a physicist, engineer and scientist.

Acknowledgements

For my ever-supportive parents. And for my friends and family. (This is what I've been doing all these years!)

Thank you for the invaluable conversations on this immense and complicated subject. In order of height: Sam Dargan, Steve Katsaros, Charles Yonts, David Sogan, Huashan Wang, Hannah Marshall and Joe Fernandez.

Thank you to everyone who supported the development of this work, including: Tania Willis, Tom Hope, Danny Kwong, Denis Williamson, Simon Collings, Glenda Thomas and Bill Acharjee.

I owe a great debt to Richard Barrett. His transformative editing made this work palatable and this book real.

Foreword

This book developed gradually during many negative years learning first hand about the problems of the developing world; and those of our so-called 'developed' societies. I am grateful, then, that the outcome is positive.

I have always been an engineer. As a child I regularly took things to pieces to find out how they worked. Twice, my practical curiosity nearly killed me: First, aged ten when I decided to wire a toy Scalextric car directly to a mains electricity plug; second, when I took the positive cable of my desk lamp and wired two copper plates into the circuit, pushing them into the springs of my chair so that when I sat down the light would come on. I still recall the tiny lightning under my control, the burn marks on the copper plate, the sparks close to my face; and the deafening bang and the flash of white light when I switched on the power to the car. This is still the closest I have come to death, yet my fascination with the invisible power of electricity was set for life. Eventually, when I could make electrical items actually *work* after my intervention, my professional life – and this book – became extensions of that same spirit of enquiry.

I first witnessed poverty at the age of 14 in Sri Lanka, the same year as Live Aid. Ten years later in Malawi, I observed at first hand the horrible lives people were forced to live. A decade further on, Live 8 and the Make Poverty History campaign were once again highlighting grotesque inequalities in our world.

In 1999, after graduating with a degree in Engineering Physics, I joined a UK solar photovoltaic (PV) company. We used PV – the direct generation of electricity from light – within rugged stand-alone electrical solutions. Ever since, I have been motivated by a desire to transfer the highly dependable 'off-grid' PV system designs of the industrial sector into the developing world.

As a specialist in stand-alone energy solutions with a focus on energy-poor regions, it was sadly very easy to see the shortcomings of projects and policies. There has been widespread bad decision making around matters of new technology and digital infrastructure. I reached a cynical point where I was articulating fundamental flaws in pretty much every energy initiative for poor people.

After wandering the world for over a decade as a single man, I am quite happy to acknowledge the cliché that it was a woman who triggered the turning point, and helped me to think more positively. There are no straightforward solutions to energy access; there are only steps in the right direction and steps in the wrong direction. Making the decisions—taking those steps— is only really possible if there is a clear end goal. That goal, in short, is *access to modern energy services for all people, organisations and nations who need them.* Those services must be dependable, affordable and locally accessible.

Discussing with others in the developing world energy sector, I became convinced that neither mains electricity, nor any kind of fuel-based or large-scale electricity network, could make this goal a reality. Worse, distributed 'solar' was steadily becoming part of the problem instead of being the universal solution in which I had placed my hopes. Amazingly, though, more people began to engage with what I was starting to articulate. I began to appreciate the many people and organisations making positive impacts – quite often with initiatives I hadn't thought of. Developing an understanding of the scale of positive potential in these organisations was key to the energy strategy I describe in these pages.

The last thing I would have predicted when I first considered writing this book was its primary focus: making profit in poor countries. During my 20 years as a specialist, the commercial value of stand-alone PV has drastically increased due to advances in component and user technologies. However, we have not made similar social advances; our generation seems unaware of the true value of reliable energy services, and insufficiently motivated for positive change.

For the first time I can see a route from the energy world we have now, to the inclusive energy access world we need. The intention of this text is to show a plausible route to the end goal: to describe some of the right steps to take, and who needs to be taking them. For the passionate people and organisations dedicating themselves to the most pressing issues of our time, I hope this book helps.

Martin Bellamy, January 2019

Summary

Selling Daylight is a realistic vision of the near future in which our use of the sun's energy can be personalised, durable, free-standing and reliable, and capable of breaking the cycle of dependency and energy poverty in the developing world.

This book is, by necessity, a long-form discussion into the highly emotive and complicated aspects of widespread energy poverty in our modern world.

The vision as narrative

If you want to understand the vision of this strategy without delving into the body of the book, read Chapter 27.

Here we narrate a typical day for a drone operator as they observe a multitude of autonomous, PV-enabled equipment that combines to describe a developing nation and its people in socio-economic progress.

Most importantly while reading, remember that everything described can be implemented now, within existing energy-related expenditure, using widely available technology and easily accessible skills.

Structure

Selling Daylight comprises seven parts with an average of four chapters each.

Part one describes the problem of 2 billion people forced to live without access to reliable modern electrical services.

It details why the prevailing solution of a mains grid electrical system – based on large central generation – has not and will not be the primary means for ending energy poverty.

Whether renewables are employed as large central generation, or as distributed supplies to support the grid, the electricity quality and dependability required for social and economic development in energy-poor nations will never be achieved without a fundamentally different approach.

Part two describes the core tenet of the *Selling Daylight* proposition: personalising energy. We justify the need to position the user at the centre of their energy services if we are to make a positive impact on global energy poverty.

We describe in detail why stand-alone PV is the generic 'hardware solution' by articulating the core economic values. Specific reference is made to how these have been neglected when applied to the developing world.

A universally accessible visual media methodology is introduced for communicating the 'knowledge sharing' requirements of our strategy.

Finally, we acknowledge the fundamental need to accept and confront the poor reputation of the generic PV format.

Part three articulates quality and high-value electrical service solutions in the context of energy-poor nations and their people.

Each major aspect of our defined requirements are then described in dedicated chapters:

- A focus on quality and value (rather than purchase cost);
- Optimising the complete holistic hardware solution;
- Propagating knowledge (using our Visual Instruction System);
- The currently damaging procurement practices and how to overcome them;
- A truly global energy services brand.

Part four has a simple core message, but one that is difficult to swallow: it is people who are the cause of poverty, and the barrier to eradicating it.

We systematically discuss why there are no monetary, technology, logistic, scale or diversity factors that we are not entirely capable of overcoming. The collective will to address the grotesque inequality of our world is the issue.

The justified solution, again, is simple but maybe uncomfortable to accept: make supplying energy services profitable for everyone concerned.

Part five looks specifically at how to apply the defining principles of *Selling Daylight* in the real world.

Specifically, we look at why large commercial manufacturers need to be engaged and how to address the present focus on risk rather than opportunity.

We articulate why the usual targeting of 'low hanging fruit' approach restricts growth and how to overcome this with versatile hardware and a dynamic commercial organisation with a long-term approach.

A successful commercial strategy requires a diversified product range from the start. Ultimately we show that revenue can come from multiple sectors, multiple applications and multiple budgets.

Finally, we look at where and how to start.

Part six justifies the essential role of local partners (people and organisations) to the success of the Selling Daylight strategy; and any initiative for eradicating energy poverty.

We start by showing how a multinational manufacturer partnered with a local organisation enables the supply of dependable services rather than mere products, and how the guaranteed reliability of these services forms the foundation for brand propagation, and with it, continued growth and profit.

We finish this section by showing how an international manufacturer working with multiple local partner networks can create the many areas of growth that lead to our 'absolutely enormous potential'.

Part seven is a reflection on the many claims of *Selling Daylight* and addresses some of the scepticism likely to result. Examples are used to convey confidence that an all-inclusive energy future is entirely possible if we accept the present realities of how our world works.

We end with a positive narrative of just what that energy-enabled future could look like for a typical developing nation.

Part two and Part three combine to define what we need to achieve.

Part five and Part six describe how a commercially sustainable business model can be implemented.

Part one and Part four are concerned with the realities of the situation in energy-poor regions. Part one focuses on the present hardware – mains grid electricity – and Part four on the human element.

Introduction

The sun

There it is, above you. The most universally recognised object to our species. Commanding awe and wonder among our ancestors; an object of mystery, of worship; the hero of this book; the source of our existence and the key to our survival.

Words like 'brightness' only undermine its gravitas. The statistics, describing the radiation and the infamous heat, do little to convey just *how much* energy the sun offers us. Our eyes, skin and hair may have evolved to deal with this immense resource. But our language, customs and society are still, in a manner of speaking, playing catch-up. A car headlamp on full beam is seemingly blinding, yet typically has around 55 watts of energy content. Compare this to the places around the world we would describe as sunny – some of which are holiday destinations, some of which suffer extreme poverty. On an average clear day, from about 2 hours after dawn for the next 8 hours, the sun delivers in the region of 1,000 watts of usable energy for *every square metre* of land. So when we say 'brightness', think of it as energy; when we say 'glare', think of it as fuel. When you have to put sunglasses on and you feel your skin burn, think of the biggest power station you can possibly imagine. Now, consider that this energy, as far as human existence is concerned, will never run out; it is highly predictable; and is absolutely free of cost.

The technology for harnessing this energy and converting it into electricity is called photovoltaics – commonly referred to as PV. PV *systems* deliver this electricity in usable formats. They have been in operation for over 40 years in the most challenging environments on Earth, as well as orbiting around Earth enabling satellites and space stations. They are the single most reliable form of electricity generation we have, and they utilise the most reliable, constant energy source we have. The International Energy Agency's *Solar Energy Perspectives* states 'In 90 minutes, enough sunlight strikes the earth to provide the entire planet's energy needs for one year' (International Energy Agency, 2011). While world consumption rises each year, the quote nevertheless serves to illustrate the vast resource available to us.

A commercial vision for positive change

The scope of current PV applications is almost comically near-sighted, most familiar in industrialised societies powering calculators or providing garden illumination. In the developing world it charges mobile phones and provides lighting. However, the true value of reliable PV-enabled services lies in sectors of both global reach and vast growth: communications, entertainment, education, healthcare and more. There is a value proposition for individuals, organisations and governments – rich and poor alike. For those without access to energy services, or those whose supply is unreliable, PV-enabled solutions can bring immense value and quality of life.

From a monetary perspective, this book outlines how to redirect spending. It describes why we need to think differently about energy provision; and how energy services, based on PV and other sustainable solutions, can be implemented within existing budgets. What are the motivations for writing such a thing?

Stand-alone PV should be a primary candidate for bringing energy services to energy-poor regions. In reality, it is overlooked due to unfortunate misconceptions, based on a poor track record and the continued presence of low quality products on the market. This book is motivated, therefore, because high-value PV is *underutilised*. This is down to a lack of understanding of how to apply the technology, as well as a perceived need to compete on cost instead of quality. When I look at the scale of need for modern energy services (over 1.3 billion people), I see the enormous potential for PV to address this need.

A second reason for writing this book concerns the complex issue of money in developing countries; in particular, the prevailing focus on *spending* rather than *value*. Working specifically with energy in the developing world for the last decade, I do not see a shortage of money. Neither do I consider the recommendations of economists, nor the commitment of funds from international bodies, to be enough. The bigger problem by far is the way *value* is perceived. This, more than anything, is preventing progress for the very people we are trying to help.

The third main reason for writing this book concerns the way we think. I wish to challenge common misconceptions of how life functions in developing nations. Often this is fed by received wisdom, derived from incomplete knowledge or sourced from biased data.

What I have seen and experienced in the developing world contrasts misleadingly with what is reported in industrialised societies. This has influenced how I approach (in this book, as in life) the intimidating subjects of poverty, energy, inequality and climate. Certainly, the challenges of communicating these issues are complex and daunting. Many have adopted an analytical assessment method at the expense of deeper meaning and context. The recent trend towards infographics is of particular concern in this regard. Statistics alone will never reflect the reality of millions of people, who are separated by continents and diverse circumstances, and who are not integrated into the electronic world. Certain solar product suppliers claim, outrageously but with apparent statistical plausibility, to be solving the world's problems. The issue lies with the rigidity and certainty by which data are used and referenced, allowing only the narrowest band of assumptions, and omitting the crucial context or timeframe in which data was obtained. Even trend data is highly misleading if the circumstances being monitored stray outside 'reasonable' limits of relevance.

Take two brief examples. Consider a statistic of 50 million low income people living in rural subsistence communities. Our assumptions of the rural poor change dramatically once those people are revealed to be in China, working in factories, exposed to health risks which their parents would never have encountered. You could describe these millions as 'newly affluent' as they are obviously no longer at risk from malnutrition. Yet smoking and fast food can be a public sign of their prosperous status. Here, in context, 'access to food' is no longer a reliable indicator of health.

Similarly, access to education, in itself, does not automatically indicate learning. In January 2014, The Education Minister of Rwanda, Vincent Biruta, admitted to the World Education Forum that while his country had achieved the highest primary school enrolment rate in Africa (with 96.5% of children in school), nevertheless, 'we have a lot to do in terms of improving quality' (Leach, 2014). Rwanda had changed its language of instruction in schools from French to English in 2009, but with a shortage of English-speaking teachers, educational achievement remains stunted.

In the same way, I feel it is important to show that 'access to energy' alone is not an indicator of capability for an individual. While it may serve some *statistical* fulfilment of need, the energy in question may well be an unreliable or sporadic electricity supply. It might take all day to gather; it might cost the day's wages or otherwise may be slowly killing those who use it. Variability of this kind is everywhere. Consider the commonly quoted metric of people 'living on less than 2 dollars a day'. This simple enough phrase conceals one of the most complicated figures to produce, where the dollar is not the native currency, where income is agricultural and seasonally variable, and where much commerce is based on bartering instead of money. Why are we concerned with accuracy? The magnitude of the number – 2.6 billion (a figure from 2008), over a third of the world's population – should be sufficient to mobilise a response. But once the numbers and infographics are printed, the challenge remains to discern its true meaning; and with it, the ingredients of a solution.

This book is a business proposition

I started writing this book as an attempt to address energy poverty and climate change using ethical and logical arguments. However, the more I worked in the field of energy services, the more I realised that my principal focus *should be a commercial one*. I believe that commercial interest alone provides sufficient motivation, for international organisations and local individuals alike, to make money supplying dependable energy services into poor regions and the wider world.

The strategy described in these pages is centred on how to make money in the autonomous energy sector. There is absolutely enormous commercial potential within the supply of distributed, stand-alone energy services. I am confident that commercial organisations must be engaged in order to see meaningful progress, and that this can be done with profit. This book describes how it can happen.

What this book is not

This book is not a statistical analysis of how much money is required for energy services, in the developing world or elsewhere. Nor do I offer opinion on exactly why so many people remain in energy poverty despite decades of effort and expenditure.

I have been careful to avoid compromising the business plan which inspired this work, and to protect any intellectual property pertaining to specific products or services. Therefore, be ready to expect a few generalisations. There is no single set path to commercial dominance; no single right answer, no 'silver bullet'. The business objective here is to demonstrate that what I say *can be done*.

In the worlds of international aid, politics, climate change, globalisation and ethics, statistical 'facts' can be made to fit with a given agenda. For this reason, I use my own basic and direct assessments of complicated principles, and I avoid quoting definitions from elsewhere. Similarly, this is not a scholarly reference, and I make very few references to academic papers or field studies.

Implementing a solution

Most non-academic analysis of the developing world comes from aid agencies or economists. This is to be expected, given the amount of international money involved. But the skills required for analysis differ wildly from those which provide answers. Where we find economists recommending diesel generators (the energy equivalent of a very high interest rate loan: cheap to implement, but expensive to operate; dependent on fuel and a pollution menace), I am compelled to step in. As an engineer, I describe the utility electrical grids of industrialised nations simply, but from a technical perspective; to raise awareness of basic realities, such as being physically disconnected from the fuels we use to produce utility electricity.

The PV industry tries to treat its primary products as if they were commodities. This is unrealistic. All PV requires technical input: PV panels generate energy depending on where and how they are used, and the electricity produced is never guaranteed, it is estimated. This technical knowledge needs communicating. With over 7,000 living languages in the world, we supply user instructions for electrical products predominantly in English, often to illiterate recipients. In these pages, we explore whether it is possible to use a single piece of media to convey at least basic knowledge to absolutely anyone.

I hope that this book can be understood by a non-technical reader who is engaged or otherwise positioned in the quest to end poverty; specifically, energy poverty. For I believe that this convergence – of the technological, commercial and ethical strands of debate – ultimately makes a compelling argument for positive change. Remember that everything that you read in this book can be implemented *now*, with existing technology, through existing supply channels and within existing expenditure.

Many issues, many solutions

There is no single solution to the big problems of poverty, inequality, fuel, climate and energy. I accept that aid organisations and international agencies represent value for developing nations, and I do not set out to criticise them. My generic observation is that the success of these programmes is dependent upon the sustained support they receive; a dependency which limits their potential for expansion. That said, certain things we *know* are not going to work: (1) don't base the energy future of nations on fossil fuels; (2) don't base them on the central generation electricity model; and (3) don't base them on any severe dependency that cannot be mitigated – including overseas aid or other conditional funding.

Stand-alone PV solutions can negate the need for large energy grid systems and can take people off of existing mains electricity. *Commercial* mechanisms are the most effective means of bringing quality energy services to regions where they presently do not exist. My strategy exists to support current pro-grammes, so that their positive impact may spread. We need to reach over 2 billion people, after all.

Who this book is for

This book is for the people who have committed to extinguish the disgraceful inequalities in our world. Its aim is to establish solar PV as a highly valu-able method of energy services provision – one that is adaptable and versatile with multiple levels of value. It is for those people in their chosen special-ist fields – education, health, agriculture, communications or commercial enterprise – who can bring their own domain-specific knowledge and utilise this technology platform for the greater good.

And this book is for you. I intend to foster an appreciation that a reli-able, stand-alone electrical service solution exists for any application which depends on energy. The goal of this book is not to try to present a single foundation to change the world. It is to provide understanding of what PV can do, of where its true value lies and how it can be used effectively by as many people as possible. It shows that there is a realistic route – which fits with how our world *is* rather than how it should be – to modern energy services for all who seek them. It will hopefully make the challenge of redistributing opportunity just a little bit easier. Switch on, and let us begin.

PART ONE

How we use energy and why change is paramount

CHAPTER 1
Electricity and people

Abstract

Four categories are used to describe the relationship of people to mains electricity in the developing world: reliable networked grids; unreliable networked grids; limited mains electricity access; and off-grid (no reasonable access to mains electricity). Each of these is explored in order to establish the principal arguments of the first section of the book.

- *Firstly, the money being spent on fuel and ineffective mains electricity in developing nations can be redirected to more effective solutions.*
- *Secondly, offsetting fuel is not, in itself, the strongest way to drive the uptake of alternative energy services.*
- *Thirdly, attracting the organisations most capable of supplying quality, cost-effective energy solutions in the volumes needed requires strong revenue potential. Such a potential exists with those already paying for unreliable electricity. (Targeting energy solutions solely at poor, off-grid communities will always be hindered by a fundamental lack of spending power.)*

> Mary has been on the phone with her sister for 40 minutes. There is hot coffee in the pot but she doesn't fancy another cup, so she takes a look in the refrigerator. She can't see what is at the back of the shelves so starts to shuffle items around with one hand.
>
> The fridge goes dark and silent. The TV commentary on the counter does likewise. The power cut is marked by a quiet descending subsonic tone, giving way to silence, as the air-conditioner, electric oven and washing machine power down. The ceiling spotlights turn off and now only sunlight illuminates the kitchen.
>
> The tablet on the side startles Mary a little when it suddenly illuminates, announcing it has lost internet connection and can no longer stream content. 'What, your power has just gone off as well?' Mary says with confusion. 'OK. I haven't got much battery left. I'll call you back when the power is up.'

The industrialised world is defined by electrical generation, and a transmission infrastructure casually referred to as 'mains', 'grid' or 'utility' electricity. To a secondary extent, some countries also enjoy gas supply through an extensive network of pipes. Since the early 20th century, big power stations and networks of cables have become an established part of the landscape – accepted for the unquestionable value they deliver to our homes, businesses and society as a whole.

Less obvious are the challenges in creating that electricity; the cost and consequences of the fuel that we use at source; the fact that we are dependent on

http://dx.doi.org/10.3362/9781788530705.001

the reliability and consistency of that electricity, as much as its plenitude. We will see how our relationship with fuel, and our attitude towards the energy that we consume, varies according to who we are and where we live. Being grid-connected doesn't simply mean being at the end of a grid cable; it also represents a standard of living, a social and cultural indicator of value. Such values may differ across the world; what is valuable in developing regions often has limited value for industrialised and affluent areas.

This non-interchangeable value is just one of the reasons why, after more than 40 years of stakeholders engaging with electricity for development, there remain some 2 billion people in our world without access to reliable electrical services.

> Being grid-connected doesn't simply mean being at the end of a grid cable; it represents a standard of living, a social and cultural indicator of value

---

Utility grid electricity

Consider the ubiquity of electrical infrastructure around us. Every electrical device has a cable – an umbilical cord to a socket on the wall. The walls themselves contain arteries which channel electricity to every room in our homes and workplaces. Some people know where the fuse board of their home or office is located; fewer would understand how this unassuming box brings the electricity from the street into our buildings in a way that makes this dangerous invisible power banal and bland. Fewer still acknowledge the conductors under our walkways; the poles erected in many streets; the cables that follow alongside roads; the pylons that stretch wires across the countryside; and the steam and smoke coming from the power stations, upon which these pylons converge.

We know that fuel goes into those power stations, but we don't usually see it. Most people will not appreciate the enormous volumes of fuels used at the heart of the transmission network. The physical infrastructure and discussion of fuel is so familiar that we simply don't notice it or pay attention most of the time.

The same is true of the product it delivers. We can plug a device into the wall and switch it on without a conscious thought. We flick a switch to light a room instantly. We press the button on the kettle and it gives us hot water. Our communications, entertainment, security, light, refrigeration, education, street lighting, traffic control, and water and sanitation – all are enabled by electricity. Even the exceptions (such as gas for cooking and fuels for transportation) are only possible because of the electricity enabling their safe logistics and management.

Meanwhile, when we are physically removed from mains electricity, we nearly always take a little bit of it with us. Batteries are a secondary attachment to mother grid; they provide us electricity for a short time before we throw them away without a thought, or recharge them by connecting a cable into the wall. We employ a multitude of inefficient strategies to take little amounts of electricity from the mains network, to consume wherever we choose.

On the grid or off the grid: four electricity areas

There are two general groups of people with respect to electricity: on-grid and off-grid. Although this distinction does not describe everything about a person's access to energy, it does offer a surprisingly clear picture of the global energy landscape. There is a world of difference between the two.

In the industrialised world, grid electricity can meet all of a person's energy needs (with the exception of transportation, though this of course is changing). It is extremely rare for buildings in urban areas not to have mains electricity. 'Off-grid' in our developed nation context refers to a lifestyle choice of a tiny number of people, and is by far the exception when it comes to domestic properties. Off-grid dwellings are cabins, holiday caravans, recreational vehicles and largely rural geographical locations. Off-grid energy services are tailored towards outdoor recreation, leisure pursuits and specific activities such as camping and boating.

In the developing world, the on-grid to off-grid distinction is less easy to define. The sheer land mass and the terrain are major factors. But when we look at the provision of grid electricity itself, the dependability and quality of the energy that empowers the industrialised world are often absent from less affluent regions. To account for geography, population dispersion and inconsistent grid quality (among other factors), I shall use the following four areas to describe the relationship of people to mains electricity in the developing world:

- Reliable networked grids
- Unreliable networked grids
- Limited mains electricity access
- Off-grid (no reasonable access to mains electricity)

Reliable networked grids

Reliable networked grids give the industrialised world the economic capabilities it enjoys. In its simplest form, an electricity supply has a single source and a single consumer linked by a single connection. In this arrangement, if either the source or the connection is disrupted, the consumer loses their supply.

A networked electricity supply enables lots of generation sources to be accessed by lots of consumers via multiple routes, so there is inherent redundancy in the system. If for whatever reason one generation source cannot meet demand, others can take up the slack. Industrialised nation electricity networks do exactly this but on a massively complex scale. (Indeed, the grid in the USA is often described as 'the most complex machine ever made'.) To meet the cost of maintaining this vast dynamic entity and to ensure reliability and profitability, the energy supply and demand is actively managed on a second by second basis; compensating or penalising various suppliers and consumers accordingly.

The *quality* of consumer electricity is essential. It is not just about preventing damage to electrical equipment but also about the stability of the network as a whole. When quality suffers, entire sections of the grid can collapse.

One instance of this is a little-appreciated aspect of energy delivery known as *reactive power* (sometimes referred to as invisible power). Reactive power is a consequence of out-of-phase voltage and current – which can be the result of the generator or consuming equipment – and is a significant factor in electricity systems. This phenomenon is believed to have caused the Northeast Blackout of 2003, affecting 50 million people in one of the largest outages in US history.

In the developing world, the term 'reliable' needs to be put into context. Official reports on energy access and quality may consider a supply with one or two outages per month as reliable. Even one outage a week during certain times of the year would be accepted by consumers without much complaint, given the largely unreliable nature of most of the grid systems.

Unreliable networked grids

London or New York, meanwhile, would not tolerate such outages. 'Unreliable networked grid' electricity is so disruptive and so damaging to productivity that it is not viable in the developed world. In this category, the network architecture is there, but plagued by a lack of reliability, with outages occurring on a daily basis, or more at certain times of the year. The problem is so widespread that even the international community has estimated those without access to reliable energy services to number in the region of 700 million people.

Physically, the network architecture contains redundancy against specific local damage or maintenance needs. This makes reliability possible in theory; however, the demands on the system and the requirements of management and maintenance are simply too costly, complex and challenging.

Poorly functioning generation equipment; badly maintained delivery infrastructure; and expensive, skilled personnel needs. These, along with the cheap nature of devices connected to the grid, are just some of the factors that result in low quality of delivered electricity. Understanding this energy sector of people – those on the grid but without the reliability – is key to appreciating a whole host of energy-related challenges, and commercial opportunities.

> The international community has estimated those without access to
> reliable energy services (in addition to those without any modern
> energy access) to number in the region of 700 million people

Limited mains electricity access

In basic terms, this category refers to the geographical and capacity limitations faced by an energy supplier attempting to control supply or stability. Where an isolated electrical demand exists (say, a factory or remote settlement), without a sufficiently high revenue potential, there is no justification to extend the grid *network*. Instead a single cable route is installed to carry electricity from the network to a remote location. These extensions are known as spurs or branches, and are common practice in industrialised and developing regions alike: the essential difference being the relative scale of the network to the spur. In somewhere like the UK, the network covers hundreds of kilometres, with a spur of typically a few tens of kilometres. Similarly, the USA network has tens of thousands of kilometres of coverage, with spurs again being very small in proportion. However, in the developing world the network can be a few tens of square kilometres – just covering a city – with spurs extending many tens or even hundreds of kilometres into suburban and rural areas.

This difference would be fine if the load at the end of the cable was known or predictable, with sufficient generation capacity on the network. Not a problem in the industrialised world: where redundancy is incorporated, spurs are sufficiently large to cater for peak demand, and with spare capacity for additional loads to be connected over time. But in the developing world these spurs are geographically few and far between, and the quality of supply is a huge problem, starting at the generation end and deteriorating in quality within the long cables the further away from the grid they stretch.

Increasingly, people migrate to these isolated lines of electricity, and try to use it for more and more services. When the magnitude and variability in demand both become unsupportable, the spur electrically collapses. Negotiating such a failure – that is, trying to power up an electricity supply with the loads that brought it down still connected – is a major challenge, as it creates a cycle of disruption.

Off grid (no reasonable access to mains electricity)

This category is self-explanatory; here, mains electricity is not a practical option for day-to-day living. The absence of mains electricity has a direct consequence upon health, economic opportunity and quality of life. The myriad initiatives to try to eradicate poverty and social injustice are largely focused on this sector.

Energy poverty

The term 'energy poverty' refers to a lack of access to modern energy services. Critically this encompasses clean cooking facilities, including boiling water; but increasingly our world is dependent more broadly on electrical *services* (of which cooking plays a part). So when we in the industrialised world talk of 'energy poverty', to a large extent we are talking about the absence of grid electricity; or more correctly, reliable, geographically accessible and stable electrical energy.

There is a fundamental difference between those in poverty and those in energy poverty: those in energy poverty often have money. For many there is of course a fine line between the two categories because those in energy poverty have to spend money on poor quality energy solutions that often leaves them with little left for other essentials. As I will describe, the core issue is not a lack of money, nor in many cases is it down to a lack of access to energy. It is the lack of access to cost-effective and quality energy services which defines those in energy poverty.

When we in the industrialised world talk of energy poverty,
to a large extent we are talking about the absence of reliable,
geographically accessible and stable electrical energy

In particular, in defining energy poverty, we need to understand the following: that *energy is different from fuel*; that providing energy is different from providing a *service*; and why one form of usage may not suit everybody.

The world does not have an energy problem

The inhabitants of the world have a *fuel* problem, not an energy problem. Nearly all of our energy originates from fossil fuels, and whether that fuel is burnt directly for heat, or burnt to obtain an alternative form of energy – mechanical, chemical or electrical – fuel is the foundation for the energy systems of our world.

And nobody wants energy anyway. Rather, we require services: heat, light, transportation, entertainment, communications and so on. It is a consequence of the Industrial Revolution and its transformation of lifestyle which has created a dependence entirely on fossil fuels. And with it, a blinkered fossil mindset.

Rich or poor, all people share this common need for services. Energy services must be reliable, available at the location they are required, safe, easy to use, affordable and as versatile as possible. Our grid gives us all of these, but only due to the sheer scale and sophistication of the networked supply; and the resources to maintain it.

But imagine if the grid failed every week or a few times a night. Or, if our appliances strangely stopped working after less than a year, instead of their expected lifespan. What if our grid was 10 times more expensive, or demanded one half of our income? What if mains electricity was only supplied to one area of town? Imagine what that would do to house prices, to security in the town as a whole. Further, imagine a wave of electricity theft in which people illegally (and dangerously) connect to overhead wires. This is the reality for hundreds of millions of people.

Energy options in the developing world

In the developing world, the presence of mains electricity can appear surprisingly widespread. But this is an illusion. Local people migrate to areas with electricity. Individuals and traders will more likely visit areas with light, phone charging, television and other electrical services. Then, businesses will locate where power is available for refrigeration, air-conditioning, internet services and so on. As a consequence, when backpackers, volunteers, non-governmental organisation (NGO) staff and business people interact with the local society, they will tend also to congregate in these electrified areas.

But in comparison to the landmass, the footprint of electricity grids is incredibly small. Some 1.3 billion people live in the *absence* of electricity infrastructure, away from the cities, suburbs and rural hubs. Include people without *reliable* electricity and the figure rises to over 2 billion. It is understandable that very few visitors actually wander to these locations, naturally choosing to stay in hotels, hostels and guesthouses with at least basic energy facilities. So there are very few people who actually *experience* the conditions under which this highly generalised category of people live.

Here, away from the mains, the basic energy requirements (for cooking and boiling water) are usually met with wood, charcoal, biomass or even manure. Kerosene is still the most commonly used fuel for lighting. These fuels are either gathered from the local region (meaning within several kilometres), or bought from local suppliers. Whether walking to purchase or scavenging the surroundings, it can take all day and is often delegated to children.

Even 10 years ago, electricity was simply not important enough to invest either the hours to reach it or the money to pay for it. At a stretch, the only common electrical service was some form of radio or music, typically enabled by disposable batteries or a vehicle battery. But since the global rise of the mobile phone, what was formerly a generic requirement for 'energy' has transformed into a *specific* need for electricity. In those instances where electricity is deemed important enough – for a local bar, shops, business or social need – the most prolific solutions by far are generators fuelled with diesel.

The real cost of fuel

The health, safety, financial and environmental costs of biomass, kerosene and diesel are well documented. But it is the underlying energy *infrastructure* of developed nations that is doing the most damage to our world and our pockets. Fuels are a major financial burden for everyone in the world, but the burden takes very different forms depending on where a person lives. Grid electricity is a low-quality solution for many energy services. It is a very inefficient way of using fuels, and is challenging to implement and manage.

In industrialised nations, we use excessive amounts of fuel compared with the services we require, but we actually pay relatively little for our grid electricity. Why is this? Subsidies and competitive market dynamics play a part, certainly; as does the maturity of our infrastructure. But more importantly, it has become ubiquitous and essential, and must be shored up at almost any cost. Our fuel dependency defines environmental, economic and political conditions around the globe.

Contrast this with the developing world. People use comparatively small amounts of fuel but they pay excessively for them, having little to no influence over international fuel supply, and little in the way of alternatives. Mains grids, where they exist, are not sufficiently reliable to become the economic or social foundation that they are in developed nations.

Wherever we may be, we all need reliable energy *services*, as opposed to fuel. One might assume that there is a common solution for all. However, does this need to be the case? Indeed, does it even follow from this premise, given the flaws in our current infrastructure, and our clear differences in affluence, geography, culture and fuel dependency?

Differing needs

When it comes to providing electrical services to the people who do not have them, we naturally try to apply what we already have to their locations. However, the strategies we employ do not fit with the demands of the developing world, where the vast majority of energy-poor populations exist. In terms of quality energy services, what might be valuable in an affluent, post-industrial society often has limited value for a rural, developing region. What we have simply doesn't fit the need.

Here, then, are the principal arguments of this section, as we proceed. Firstly, the money being spent on fuel and ineffective mains electricity in developing nations can be redirected to more effective solutions. This concept is already being applied in specific areas such as solar PV lighting to replace kerosene lamps, but so far only in the tens of dollars that individuals spend on fuel rather than the tens of millions in fuel expenditure by commerce, governments and international bodies.

> The energy strategies we employ do not fit
> with the demands of the developing world

Secondly, offsetting fuel is not, in itself, the strongest way to drive the uptake of alternative energy services. For the huge and growing sector of electricity consumers stuck with unreliable grids, it is the *dependability* of those services that will hold immediate value.

Targeting energy solutions solely at poor, off-grid communities will always be hindered by a fundamental lack of spending power. Attracting the might of the multinational energy service suppliers (the organisations most capable of supplying quality, cost-effective solutions in the volumes needed) requires strong revenue potential, not good intentions. Such a potential exists with those paying for *unreliable* electricity, and it can be used to establish and propagate local energy services within reach of those who most need them.

When quality, reliable and affordable energy services become available, within the budgets already being spent by individuals and nations, we will have a foundation to enable positive change for billions of people. We will also have a foundation for retaining our own standard of living into the future without the present dependency on finite fossil fuels. Both cases promise enormous commercial potential. The strategy described in this book is all about highlighting that commercial potential, and how to supply valuable energy services solutions in response to present and future demand.

CHAPTER 2

Fuel and our lack of awareness

Abstract

The electrical utility grids of industrial nations are structured around, and dependent upon, fuels extracted from the ground; and they are woefully inefficient. Upwards of 75% of the precious fuel resources that we feed into the grid electricity system is lost. Yet we are propagating this model into the developing world – where the consequences are even more severe.

The policies of the industrialised world are making the problems of global pollution and fuel dependency worse, and instead of solving the energy access challenges of the developing world, we are actually prolonging them.

Excessive fuel use, wastage and global pollution; you would think these would persuade consumers to change their ways. But at best they only feature as supporting arguments when trying to justify alternative energy services. Even the financial and social arguments are marginal issues. A different approach is needed.

The people who depend on fuel the most are those who care the least about it.

Many of us are bound within industrialised societies which rely for their electricity and transport on fossil fuels – most of which we waste. Our consumption creates massively more pollution than the resultant services justify. We do this while funding some of the most ethically questionable and highest polluting organisations in the world. Despite the rhetoric of recent decades about clean energy and lower carbon emissions, our global impact on the planet is getting worse, not better.

Electricity generation accounts for around one-third of global pollution. Yet the people who collectively spend the most money on electricity – and therefore have the strongest commercial leverage – are those who have the lowest awareness of fuel or its impact. Conversely, those who spend the highest *percentage* of their low earnings – on unreliable and low-quality electricity – have comparatively high fuel awareness, but little commercial influence.

Our disconnection from fuel

There are two main reasons for our continued apathy towards fuel dependency. The first is that we are so disconnected, emotionally and physically, from the fuel itself that it is no longer central to our personal day-to-day lives. The second, related factor is our lack of understanding of energy services, in particular how fuel is used to produce electricity.

http://dx.doi.org/10.3362/9781788530705.002

Those of us who live in industrialised nations have little direct interaction with fuel. Specifically, I mean for lighting, heating and cooking, where we purchase and handle the fuel ourselves. When cooking with gas, we are unlikely to consider the cost of the gas coming out of the stove, where it came from, or the infrastructure behind its delivery. We are most likely just thinking about the food it is cooking. As recently as the late 1950s, when my parents grew up, people still used coal for household heating and cooking (partly the cause of the infamous London smog of 1952). The coal was delivered in sacks that had to be personally paid for and stored about the premises. People had to physically lift and load coal into the burner. Some households still use coal, albeit in a much cleaner and more orderly manner. More common today is wood; and the people I know who have open fires or wood burners are acutely aware of the effort required to source, store, move and burn it. When visiting friends or family who use wood for heating it is standard practice to wear warm clothes until the falling temperature justifies the effort of lighting a fire. When that fire is lit, doors and curtains are kept closed to keep the heat in – and children are shouted at when they let the heat out. A practical, self-imposed household energy efficiency policy.

We are so disconnected from fuel itself that it is
no longer central to our personal day-to-day lives

But many of us with central heating just set the thermostat and forget about it. We might turn the temperature up on especially cold days, or if we happen to be in the house when the heating is not scheduled to be on; but mostly we don't have to think about it. We sit in a T-shirt and heat the room accordingly. We heat the whole house even though we are only using one room at a time. This also happens in hot climates, when people wear insulating clothing in chilled, air-conditioned buildings.

Central generation

'Central generation' is the term used to describe the standard utility grid arrangement in all industrialised nations. The term derives from the relative scale of generation to consumption with respect to capacity and quantity. A national grid can have just a dozen or so power stations capable of supplying millions of consumers. Even a single large power station can supply the electricity demands of thousands of consumers many kilometres away in all directions.

It allows large amounts of fuel to be processed at a small number of sites, rather than at the premises of each user. This of course is far more attractive for consumers and leads to cleaner residential environments. In theory (if not in practice) it also allows pollution to be controlled centrally.

The utility grids of industrial nations are structured around, and dependent upon, fuels extracted from the ground. Since large, central generating power stations were first commissioned in the late 1800s, the percentage mix of coal, oil, gas and nuclear fuels has varied over the decades, affected by economics, availability, political allegiances, natural disasters and, sometimes, even environmental factors.

Historically, the last few decades had seen gas overtake coal as the most widely utilised fuel, being both cleaner and easier logistically. (The exceptions, France and South Korea, are primarily supplied by nuclear generation.) More recently, however, coal has experienced the greatest growth in demand, particularly from China and other rapidly growing nations who simply cannot obtain sufficient alternatives. Coal is now the biggest energy source for electricity generation in the world because it is *generally* the cheapest in terms of purchase cost and construction of power stations to burn it. Despite coal being the dirtiest fuel we have, over the past 30 years the amount of electricity produced from coal globally has tripled.

I say 'generally' because the global fuel sector is in a constant state of flux. Global politics play a defining role in fuel sourcing, and variations in fuel prices impact entire nations. Meanwhile the demands of utility grids – and therefore their fuel requirements – vary throughout the year, so constant adjustments must be made to ensure an uninterrupted supply of the precious commodity – which, mostly, we go on to waste.

National grids: fuel dependence and waste

Our national grid generation and transmission infrastructure is woefully inefficient. Power stations receive fuel which has often undergone extensive processing and travelled hundreds or thousands of kilometres. This fuel is burnt (in the case of coal, gas and oil) or manipulated (nuclear) to produce steam, which is used to spin an electrical generator. This is an inherently inefficient process. By the time any actual electricity leaves the power station, over 50% of the energy content of the original fuel has been lost.

The electricity is then conducted over tens or hundreds of kilometres to regional sub-stations. This network of transformers converts the electricity into the standard format compatible with our appliances, then supplies it to our homes and businesses. At this stage, anywhere between 5% and 25% of the original power station fuel-energy is lost from transmission.

Thereafter, we waste a great deal of our electricity through inefficient use. Any unintentional heat or noise from an appliance is wasted energy. An incandescent light bulb could be using less than 5% of the raw fuel energy content to produce light. (A light bulb is too hot to touch after only a few minutes; all that heat energy has to come from somewhere.) If only 5% of the initial energy ends up as light, that means 95% of the energy content of the raw fuel fed into the power station was wasted or lost to inefficiency. To add insult to injury, consider that fully 10% of total USA and EU electricity consumption is accounted for by appliances on standby, doing nothing. (As an aside, the average car is no better, converting around 20% of the fuel energy we put into the tank into actual motion. Any unintentional heat, vibration or noise is just energy being wasted.)

In all, fuel energy wastage varies from region to region (due to fuel type, age of the infrastructure and the distance of transmission), but we throw away upwards of 75% of the precious fuel resources that we feed into the grid electricity system.

Just consider the severity of that figure for a moment. Think of the effort, cost and resources we expend in obtaining coal, oil, gas and nuclear fuels, only to waste three-quarters of that energy potential.

The percentages for each stage of inefficiency will vary from region to region, depending on what is being powered and where. In fact, the figures will vary depending on the time of day. Some energy networks will be better than others. However, the numbers I have used are entirely representative of grid systems around the world. Meanwhile, note that none of these examples includes the energy cost of extracting, processing and delivering the fuel to the power station in the first place.

We throw away upwards of 75% of the precious fuel
resources that we feed into the grid electricity system

People new to grid electricity

This prized grid electricity carries an unfortunate consequence when it reaches individuals who grew up without it (or those who, at least, previously experienced using fuel at first hand) – it often restricts them from practising the energy efficiency they learned when they were younger. When people from un-electrified rural areas migrate to cities, they are forced to adopt high-energy consumption practices. They cannot conveniently use a fire to heat water in their tower block flat, so they use a kettle. That kettle runs on electricity generated many kilometres away using relatively large amounts of fuel. The same is true for air-conditioning. Where they once would have used fans, opened doors and windows, or generally worn less, they now keep the windows shut because of the noise outside (as well as the pollution). They would also prefer to be cool in their work clothes.

Cooking in a tower block flat is nearly always electrical, itself a significant forced leap in overall fuel consumption compared with direct fuel burning. But if you have an electric hot water shower where before you only had cold water, of course you are going to use it; and for longer, more regularly. Producing heat from electricity is an intrinsically inefficient process.

People new to electricity grids do not simply represent a shift in energy demand from one fuel to another. They herald massive increases in overall fuel consumption; even if they remain as energy-aware or as vigilant as when they handled the fuel themselves. Even if they still wear warm clothes and only heat a single room in cold weather, or remain frugal with air-conditioning in hot climates, that electricity requires several times more fuel to perform the task than if they had done it themselves.

Grid electricity brings with it an aspirational lifestyle, where consumption is synonymous with growth. People new to utility electricity wish for more than climate control for their homes; they want the TV, fridge, lights, stereo system and all the trappings of the affluent nations. Those new to the grid mostly don't have any background or general awareness of electrical energy efficiency (as opposed to raw fuel efficiency), and these principles are not taught. This lack of understanding just exacerbates the problem.

When people from un-electrified rural areas migrate to cities,
they are forced to adopt high fuel consumption practices

This reality is why so many new grids in developing nations do not function correctly. When supplying new populations with electricity, the load requirements are calculated based on modest energy demands for the established regional population, with conservative spare capacity on both fronts. But once those people are finally connected, the load proves far higher than planned. In addition, people migrate to the electricity network lines, just as people used to settle on rivers, rail routes and then major roadways.

Failure on demand

As more and more people connect more and more appliances, the demand continues to rise; and something has to give. The first victim is the *quality* of the electricity: technically speaking, this means consistent voltage and frequency – vital for efficiency and correct function of equipment. Eventually, during heavy load periods (such as very hot days when air-conditioners are needed; or, at meal times), the mains grids regularly cut off. This can happen seven or eight times in an evening in some African countries I have worked in.

Variations in the quality of electricity (influenced by excessive localised demands) and failure of grid sections all lead to even lower operational efficiencies – and higher fuel use. Millions of new homes, particularly those in emerging economies such as China, India and Brazil, are entirely electric. I'm not suggesting we should encourage people to burn wood or coal in their homes instead of electricity, or live with the windows open all the time, but the severity of moving people who were not on the grid and limiting them *exclusively* to the grid, places a disproportionate additional burden on the infrastructure and fuel resources.

The burden of fuel

For the comfortable citizens of the industrialised world, our fuel burden is in the background, unobtrusive yet integrated into every aspect of our lives. We don't get our hands dirty, we just work the extra hours to cover the cost and accept commercial, political and societal allegiances in order to maintain the wasteful energy supply systems.

At the other end of the energy spectrum, for developing world inhabitants who use fuel directly, the fuel burdens are cost, time, limited versatility, health and safety. Fuels are a notable proportion of many household incomes; they can take many hours to acquire, they can only be burnt directly or they can only be used for small electrical devices. Low quality kerosene lamps can produce as much smoke as light; while the health implications (and self-evident safety issues) of burning fuel indoors for cooking and boiling water cause even greater levels of suffering.

In between are the newly electrified citizens of emerging economies, who represent a massive increase in national and global fuel use. With this increase comes the international dependency that keeps those nations in the very state from which they are attempting to escape; overshadowed throughout by the global environmental effect of burning all of these fossil fuels, as more people move from personal fuel use to grid consumption.

Excessive fuel use, wastage and global pollution; you would think these would persuade consumers to change their ways. But at best they only feature as supporting arguments when trying to justify alternative energy services. Even the financial and social arguments are marginal issues. The reality is that you will find greatest fuel inefficiencies where populations have the least understanding of fuels and energy. And in general, where they are least motivated to change for the better.

What is electricity anyway?

The greatest impact of central generation is the apathy it has created among its users; to the extent that we are largely ignorant of the commodity we consume. Electricity is the single most important technology of our societies, yet the vast majority of electricity consumers could not explain what electricity is, how it is made or even how much they use. We pay the bill when it comes through the door (pausing to moan about the cost), or by direct debit, hidden from view; and we continue regardless. Why do we know so little about something on which we are so dependent?

Disassociated, distracted or disingenuous?

We don't understand what electricity is because there is no effort in obtaining it; and no significant impact from using it. Everything just works, grid electricity does what we need it to do, and it is everywhere we need it to be. We don't have dirty fuels around our homes or the smoke and toxic gases in our faces, so we just don't think about them. As long as the supply is not interrupted, we put the consequences of its delivery out of mind.

This is nothing new. Distracted by the comforts of modern life, we conduct similar relationships with food (ignoring the horrific cruelty imposed on mutated animals for the supply of our meat), with clothing (awful labour conditions and working hours imposed on oppressed foreigners so we can have cheap garments), with electronics (toxic and unsafe environments) and of course with our wealth in general (at the expense of a billion in poverty and oppression).

The contradictions we endure in order to continue our lifestyles are a direct and ugly consequence of its benefits; and we try our best to ignore them.

What else contributes to this void of curiosity? The absence of fuel about our homes (and the basic learning it once instilled) is a significant factor. The lack of formal education is another.

Education, education, education

In my first year of electrical and electronic engineering study in the late 1980s, we were told of the big push to bring engineering education into the mainstream. The justification was simple: our society is now dependent on electricity and we have a shortage of people who know what it is. This was before the internet, before mobile communications and before global e-commerce. Today, we still don't teach electricity properly in schools.

> We don't understand what electricity is because there is no effort in obtaining it; and no significant impact from using it

Those of us who grew up since the 1970s have had little or no experience of handling fuel; we have only really known electricity and piped gas. Because there is so little interaction with the functionality of these systems, we have never been incentivised to understand them.

Many argue that we don't need to know. But we do. Fuel awareness and electricity generation should be taught in high school. Not necessarily the deep technical theory, but rather the implications of having it: an appreciation of the raw fuels; the processes for converting them into electricity; and the networks through which it reaches us. We need to be taught which electrical services consume the most fuel. This awareness would give learning energy efficiency a central purpose – a quantified meaning, if you will. Professional and government organisations are still committed to increasing the profile of engineering in schools and in society as a whole. But after decades of trying, and despite our dependency on electrical services being more acute than ever, science and technology are still marginal subjects. Many engineers have just given up the push. Like them, I think that people will appreciate how integral dependable electricity is in their lives only when it is no longer there.

Pollution

Mains electricity generation models separate the consumer from the fuel and the level of system efficiency; this, if you'll forgive the pun, is its central problem. They also remove the decision-making ability of the consumer regarding the origin of the energy. The majority of electricity consumers have little choice of what fuels are used, or how pollution from the burning process might be managed, or mismanaged. (Thankfully this situation is improving, with companies such as Good Energy in the UK offering electricity from 100% renewable sources.)

This lack of fuel choice happens at the supplier end as well as the consumption end (power stations are fuel-specific). Meanwhile our societies have regulations to ensure that our air and water are just about clean enough (although many countries are found in violation of these legal environmental limits). This conveniently contributes to our aloofness (or, perhaps, to our abdication of responsibility) regarding both localised and more widespread forms of pollution.

When I started to learn about energy generation and fuel processing, one thing never failed to surprise me. People who live in nations which have been industrialised for many decades – people who once suffered the severest pollution in their homes and on the streets of their cities – have so quickly forgotten what happens when you burn biomass, wood, coal, oil and gas. The glossy technological cover to our functioning world betrays the effluent which festers behind. To witness the real world of fuel and pollution you don't have to look into the pages of history. Just visit a developing nation. At the other end of the social comfort scale, people handle fuel every day; pollution is literally in the face of the user, instead of being globally diffused and hidden from our high-density population centres; hidden out of sight and out of mind.

Fuels in energy-poor countries

But people in developing countries often have no alternatives to using fuel, and little power to influence change; a fact demonstrated over the many decades that we, the empowered, have known about their struggles.

In rural communities, individuals (mostly women and particularly children) spend hours each day finding or purchasing raw fuels – biomass, wood or charcoal. These have horrible consequences for pretty much everyone; burnt on poorly constructed indoor stoves, resulting in a whole range of damaging health effects, and of course death. Kerosene alone produces toxic fumes equivalent to the intake of an average chain-smoker. Two-thirds of adult female lung-cancer victims in developing nations are non-smokers.

There is no sweeping alternative to this. Fuels remain essential for cooking and boiling water, especially as water-borne diseases pose an even greater menace than pollution. While clean-burn stoves *are* being used, as well as fuel alternatives such as solar lights, these are still scarce compared with the enormity of the need.

When referring to *total energy poverty* – including all fuels and electricity – a figure of 2.5 billion is the number to sit back in our comfortable chairs and consider. This is the number of people who rely on wood, charcoal, biomass, kerosene, diesel (and anything else they can get their hands on) in the absence of 'modern energy services' – a significant part of which is reliable mains electricity that does everything for those lucky enough to have it.

> 2.5 billion people rely on wood charcoal, biomass, kerosene, diesel and anything else they can get their hands on in the absence of 'modern energy services'

Pollution in poor cities

It is not just those in the developing world without mains electricity who suffer. Even in electrified cities the pollution levels are unbearable for anybody accustomed to a controlled environment. Economically struggling nations have either badly enforced regulations for fuel use and pollution, or none at all. Technologies to control pollution are either dysfunctional or financially out of reach. The result is highly localised pollution from vehicles and properties, along with less obvious environmental oppression by the accumulation of toxic emissions. This is especially true in densely populated areas.

Fuel dependency and pollution are both personal and national. They are economically and socially restrictive in the extreme. Ordinary people are very aware of the consequences of burning fuel, especially when it spills out as thick black smoke from passing vehicles. And those within the energy industries and environmental sectors know that poor operational efficiency in their power stations and transmission infrastructure leads to poor quality electricity, as well as poor national and international environments. But these people have relatively few options, so they look to the industrialised nations for guidance and help. And this is where the real problems lie.

Changing our motivations

We humans face an unfortunate reality when it comes to our motivations and our ability to change things for the better. Reliable networked electricity has enabled prosperity and a high standard of living in the industrialised nations. But such centralised utility serves only to disassociate users from the very systems on which they depend; while simultaneously reducing our understanding of them. Consequently, it acts to suppress our motivations to seek alternatives.

Conversely, those off the grid – the energy-poor – have all the more motivation to reduce fuel use, because they experience their dependency with a daily immediacy. Unfortunately, they lack the prosperity to develop alternatives, or influence others to develop them. In short, those able to change fuel dependency are not motivated; while those motivated are not able.

For at least two decades, international movements and initiatives have been trying in vain to incentivise wealthy nations to change their ways. Those incentives have been as vivid and as compelling as you can get: the threat of global climate change and environmental devastation; the rising cost of fossil fuels to the point of economic unsustainability; and the consequences of funding overseas regimes who supply us with our fuel but who also nurture forces which seek to destroy our very way of life.

But we choose to remain ignorant of the circumstances of our comfort. It is unlikely that we in the privileged world will be motivated to change by a logical argument, or by the very *real* threat of something that *might* happen.

It is the *dependability* of the grid, more than its physical presence, that underwrites our modern lives. Many industrialised nations are dangerously close to losing the absolute reliability that we take for granted, and in the developing world dependable utility electricity is always going to be out of reach. This, however, is a complicated reality to convey to those who decide energy policy for both rich and poor nations. Our lack of education is twofold, failing to grasp either the complexities of utility grid electricity for ourselves, or the challenges of trying to provide it in developing nations.

Changing our own practices has proven a monumental task. But we continue to push our misguided prevailing wisdom onto others; and that needs to stop. We are making the problems of global pollution and fuel dependency worse, and instead of solving the energy access challenges of the developing world, we are actually prolonging them.

We will never be appropriately motivated to respect our energy services until the day we are required to make an effort to obtain them, or if we become personally inconvenienced by the service itself. The developing world experiences both of these realities: the former where utility grids are not present; the latter from the grids.

The industrialised world does not have either. The driving principle of our modern societies is to make things easier for us, so it is unlikely we will place a tangible human value on electricity anytime soon. But the hangover from this party could be closer than many people think.

> Many industrialised nations are dangerously close to
> losing the absolute reliability that we take for granted

We must accept that the environmental damage from our fuel consumption, the financial burdens and even the threat of societal disruption are not sufficiently strong influences for us to alter our habits. As long as we ignore how fuel is used for our electricity and transport, how it is extracted or even know where it comes from, it will not become a visible limiting factor for our civilisation. But something else might.

CHAPTER 3

The developing world cannot have reliable grid electricity

Abstract

It is the dependability *of the grid more than its physical presence that underwrites our modern lives in industrialised nations. This high quality and reliability cannot be implemented across the developing world. Therefore, neither can the social and economic foundation that it represents.*

We explore the many factors behind the dependability of industrialised nation grids, and why these cannot be transferred and implemented into developing nations.

Even upgrading the grid capacity with central and distributed generation, there is a simple reason why this will not serve to fix the mains electrical systems in the developing world: there are too many people wanting electricity.

The problems of the developing world are becoming worse – at a worryingly fast rate – because of the shortcomings of the grid format.

Electricity networks are probably the most significant threat to our standard of living. These mammoth systems are decrepit, and restrictive to modernisation; our demands ever higher and more exacting. Initiatives for change – to build more power stations, replace transmission lines, upgrade switchgear and so on – often fly in the face of carbon emissions reduction commitments, overseas fuel dependencies and economic austerity. Our disconnection from fuel and our lack of understanding have led us to a critical point of vulnerability, which is causing severe unrest behind the glossy façade of our energy suppliers and governments. Something will eventually have to give.

That said, the cracks may appear slowly at first. We are likely to lose electricity for minutes, rather than for days at a time. Later, fairly regular if brief periods without electricity may give way to outages of a few hours. It will be interesting to see if patterns emerge where we lose the supply at regular intervals; and how these interruptions are communicated or explained.

Keeping the lights on

In recent years, operational grid failures (meaning those not due to weather or other natural phenomena) have affected entire cities, or tens of millions at a time, in so-called advanced economies such as the USA, Italy and Australia. Meanwhile the biggest grid failure blackout in history hit hundreds of millions in India in 2012.

http://dx.doi.org/10.3362/9781788530705.003

NEW DELHI – India's energy crisis cascaded over half the country on Tuesday (31 July 2012) when three of its regional grids collapsed, leaving 620 million people without government-supplied electricity for several hours in, by far, the world's biggest blackout.

India's demand for electricity has soared along with its economy in recent years, but utilities have been unable to meet the growing needs. India's Central Electricity Authority reported power deficits of more than 8% in recent months (Associated Press, 2012).

Consider for a moment that 620 million people is about twice the population of the USA. Hundreds of grid failures occur every year. Very few are caused by insufficient supply capacity; the problem lies in how the infrastructure branches out from relatively few power stations to millions of consumers.

But these failures have failed to motivate us to action. With every year of delay, they become more severe and more expensive to remedy. Once they become more frequent, they will reach a critical point of economic and personal impact; and there will be a rush to try to fix the unfixable. Industrialised societies are enabled by the *reliability* of electricity. Just a few seconds of interruption can massively disrupt our economies and our everyday lives; while regular power outages of an hour or so would threaten to undermine our whole way of life.

It is not just the duration of power outages that causes the disruption; it is their timing and their frequency; and the only way to appreciate this is to experience them directly. I am not suggesting people deliberately shut off electricity to their homes or offices. Instead, consider a few scenarios.

- A man walks into a room of people watching the final of a sporting event (a football match for example) and switches off the TV just at a crucial and decisive moment.
- All the computers at work crash for the fifth time that day. That deadline you were hitting is beginning to look less realistic; and the disruption, until now an annoyance, is beginning to show financially.
- You are walking through an unfamiliar city at night and all the lights suddenly go out.

Sure, assuming that power was available more often than not, we could continue to cope with regular disruptions. But if we could not rely on lighting, elevators, computers, internet access or electronic banking (for starters), our confidence or ability to plan, even in the short term, would suffer. Our economies would slow, our productivity would reduce and our standard of living would fall.

Existing grids and reliability

There are many ways to make grid networks more stable in industrialised nations. You can physically upgrade them to make them more efficient; you can promote energy efficiency at the demand end (which has a magnified benefit given the inefficiencies of supply power); you could introduce sophisticated energy management systems to redirect power to demand centres in a matter of minutes; and you could of course add new generation capacity. The last of these is the most commonly proposed option for addressing reliability in developing nations. This is partly due to the social or technical challenges posed by the other options, but also because it brings proven benefits to industrialised national grid networks.

Distributed generation

Furthermore, you can connect electrical generators of smaller capacities than the primary power stations, but locate them strategically across the grid network, closer to the areas of consumption. This is known as distributed (or 'decentralised') generation. A solar array on a residential house for example, is as close to the point of consumption as you can get.

Distributed generation is a way of stabilising electricity quality. Using advanced monitoring and real-time demand response, it helps the grid to cope at peak times of the day; or with concentrated consumption, such as major televised events (your royal weddings, Super Bowls and talent show finals). In industrialised nations, this advanced form of distributed generation proves very effective. Technically, making the 'smart grid' work is not especially challenging. The politics and economics are another matter altogether. There are constant scaremonger stories about how the grid cannot cope with too much solar, wind or other methods of generation – that is, methods which are not already controlled by the existing stakeholders.

A solar array on a residential house is as close
to the point of consumption as you can get

But there is a simple reason why distributed generation will not serve to fix the mains electrical systems in the developing world: there are too many people wanting electricity.

More consumers, like it or not

Industrialised populations are already established on the grid. The population does not grow very fast, and the number of users remains stable. Energy efficiency regulations help to balance increased usage in specific areas, while helping to address the reduced effectiveness of the ageing infrastructure. In the UK, for example, the whole population is served with electricity. By contrast, Kenya has over 20 million people who do not have reliable electricity and want the grid; while Nigeria has upwards of 75 million. Grid reliability is not dictated by numbers alone. More relevant is how much the actual demand outstrips the population for which the grid was designed; when, inevitably, those additional citizens connect to it. The 20 million in Kenya (incidentally one of the more advanced African countries from an infrastructure perspective) represent over 40% of the population; for Nigeria it is nearly 60%, which is by no means unique. Of the 105 million Ethiopia population, nearly 60% presently do not have the grid.[1]

How do you improve stability and reliability, or serve more people? You could feed more energy into an existing grid, but this is a stopgap at best. A Band-Aid. It will only last until the number of users reaches a critical level, overwhelming the supply capacity and dragging it down to its present unreliable state.

Unfortunately, this is exactly what international organisations and government programmes are doing. Numerous central generation projects costing billions of dollars are being implemented. All of them, by definition, will underperform and fail to deliver true value. More generation will not work.

The scale of the land

The central generation model is limited by a simple matter of geography. The sheer size and terrain of many developing nations rules out a networked grid system, even with long spurs electrically supported at both ends. To put somewhere like Africa into perspective, the USA would comfortably fit three times into our world's poorest continent. In fact, the African land mass is larger than the USA, China, India, Japan and the whole of Europe combined. (Take a look at the great map titled 'The True Size of Africa' easily found by searching for 'Immappancy', by Kai Krause (Krause, 2010).)

Realities of the grid

Even with the greatest diligence and expertise available to design new central generation systems (or to modernise existing networks), there remain very practical, human reasons why we should not do so.

Mains electricity systems are critically dependent on skilled personnel for safety and stable operation. Some are based at the power stations, or centred on major electrical infrastructure; but it is just as important to have technicians and engineers to respond to localised problems on the transmission network.

In industrialised nations, with tens of thousands of engineers dispersed among the population, we are only just able to meet the hardware and supply challenges of our grids. When homes and businesses lose electricity due to storms and other natural phenomena, the duration of the blackout depends on how quickly skilled engineers can identify and reach the cause. In most cases this will be a cable failure of some kind; from a tree falling on it, excessive thermal stress or compromised structures from flooding, for example. We should all be appreciative of the skill required to fix such dangerous electrical faults, often in the middle of the bad weather that caused them.

The central generation model is limited by a simple matter of geography

---

This level of responsive expertise is simply not available in developing nations; nor is it economic or practical to implement. Any high voltage (mains level)

electricity requires skilled personnel, located at the place of generation and able to respond to operational challenges throughout the distribution network on an hour-by-hour basis. Further afield, even establishing smaller, more localised mains grids (or 'mini-grids') requires industrial installations with formidable maintenance and safety demands.

Trying to attain grid reliability in this context – by creating a specialist, highly trained *human* infrastructure, with the necessary (and expensive) tools and materials essential to their work, over such vast distances – is simply not practical. Even if sufficient money was available, the skills are not. And in the absence of reliability, the value of the services it enables is critically undermined. Such a grid would not justify even its running costs, let alone the central fuel cost.

People migrate to the grid

People have always migrated to cities. Central generation is logically based around urban centres, where sufficient revenue exists. But in energy-poor nations, spurs extend outwards much further from the central grid than they should, serving marginal communities where revenue is limited. But people are naturally attracted to the mains electricity and everything it promises.

By their very nature, existing utility networks in developing nations are unreliable. Putting aside the many environmental factors (heat, extreme weather, floods and droughts) and no matter how well designed and maintained they are at a given moment, there are millions more consumers waiting to find a way to tap into them, legally or otherwise, and pull that reliability to the floor. Inevitably migration overloads grids, at everyone's expense; this will always be the way with grid systems in these countries.

Electricity and prosperity

For those aspiring to the developed world's standard of living, the relationship between widely available electricity grid services and prosperity is extensively documented, and compelling. It is reasonable enough for them to conclude that electrical utility infrastructure is the foundation for economic and social prosperity. But this is only partly true. Mains electricity without reliability or dependability is a very expensive and damaging use of fuel, which limits economies, productivity and living standards. Without dependable lighting, computers, internet, communication, healthcare, education, security, safety and even entertainment, the foundation for economic strength and progress becomes seriously unstable.

It is reasonable enough to conclude that electrical
utility infrastructure is the foundation for economic
and social prosperity. But this is only partly true

Reliability in networked grids

For all the hazards of a top-heavy, creaky, over-subscribed utility infrastructure, we in the industrialised world just about manage to get by. Our grid, despite everything, remains reliable and dependable, for many reasons. Here are just a few.

- The generation and transmission infrastructure has been systematically developed and expanded for over 100 years to meet our needs and our locations.
- As the scale and complexity has increased, so have skills for management and control of these monster networks.
- As electricity has become ever more critical to our standard of living and our economies, a vast array of safety, security and management technologies have been applied to the grid, all with specialised engineering resources to support them.
- Electricity became highly profitable for fuel suppliers, power station operators, network operators and the thousands of component suppliers. The electricity industry is a global, commercially competitive market – with multiple options for every part of the system.
- Governments underwrite the operation and the reliability of their national electricity systems.
- We can afford both the fuel and the cost of maintaining the infrastructure, which converts it into electricity and delivers it to where we want it.
- We live in countries which are sufficiently influential or wealthy to ensure an ongoing fuel supply – albeit at a significant price.
- We have (and enforce) quality standards for equipment connected to the grid; so a factory operating, say, large-scale motors will not have a detrimental effect on the nearby residential supply.
- The vast majority of users are connected to the *network*, not a single node protruding from the power station. So there is redundancy in their supply.
- Our grid networks are dynamic; supply can respond to changes in demand on a seasonal and a minute-by-minute basis.
- We have national agreements for importing and exporting energy either for normal operation or for unexpected high demand or low in-country generation.
- Our grids are secure and safe. They cannot simply be shut off by a single dictator or political body as a means of population control or punishment.

Developing nations cannot meet many, if any, of these commitments long-term. Notice that most of the points above are a consequence of longevity, robustness, influence or affluence. These are emergent qualities of developed nations; the operational maturity of the infrastructure (as opposed to its mere *age*), and the social or political stability of the host nation. These are unlikely to be the case for nations which are at an earlier stage of economic capability.

We cannot magically transfer reliability to the developing world. The dependability that is so essential to our modern lives, and where the value of grid electricity really lies, itself depends on the characteristics of established, affluent nations. As with so many aspects of our comfortable lives, we cannot just transplant a utility electricity model into developing nations. Our grid cannot be their grid.

So what to do? At the very least we should stop making the problems worse. We need to stop building central generation grid systems in places that do not have the ability to manage or maintain them. How do we go about such a challenge, when the stakeholders of energy in the so-called developed nations are so determined to replicate these solutions wholesale?

The developing world may not have much to be pleased about when it comes to energy, but it does have one advantage over the rest of the world: its potential. Opportunities for tailored, low-fuel-dependency electrical services represent a real and credible option, in part because the existing 'solutions' already visibly cause widespread personal inconvenience, personal cost and personal loss.

Harming the developing world

I have gone into some detail about grids here, because it is impossible to have a grounded discussion about energy service provision without first discrediting the default. Centrally generated utility electricity is the prevailing model, against which all other forms of electrical energy provision are compared. But it brings with it a great deal of bias and lack of understanding, which is seriously hindering societal and environmental progress.

That grid electricity is inappropriate for developing nations is widely understood by anyone familiar with poor countries and the challenges facing their people; but there is disagreement as to the extent of the statement. Most accept that rural areas will never have grid electricity cables extended to their locations. But many assume suburban areas will eventually, inevitably acquire it, while cities will *always* need utility electricity.

I do not agree; there are better options. In its heyday, the grid was a marvel of engineering with no equal to the services it enabled. But times change: technology has advanced to the point where there are proven alternatives to mains electricity that, eventually, will become the default. The grid is no longer the best solution for anyone – no matter where they live in the world.

The grid will not be superseded because of a greener or cheaper alternative. It will be surpassed by something that is, simply, better. Specifically, I mean by something that fits our immediate needs and lifestyles, is more financially attractive as a whole, and releases us from the natural and human threats posed by dependency on fuel and large infrastructure models. Crucially, the best energy service solutions will be the ones we own ourselves, can personalise and can utilise as we like, when we like. The grid falls hopelessly short of this ideal.

Shortcomings of the grid: a summary

- The utility electricity format is dependent on, and highly wasteful of, fuels which cause environmental damage through extraction, processing, burning or disposal.
- High capital funding is needed for all aspects of networked grids – from fuel supply to the electrical infrastructure.
- Large skilled workforces are required across the network; specialist tools and training are essential; they must continually repair and upgrade the equipment just to keep it operating.
- Daily operation itself is highly complex; it requires minute-by-minute adjustments in generation-versus-consumption balancing. The grid can only operate if that balance allows the quality of the electricity to be maintained.
- Utility grids create dependency for the host nation and for its citizens, whether they are connected or not. Their scale and complexity alienate consumers from the process of delivery.
- In the developing world, the dependability of mains-powered electrical services is absent – and so their very value is fundamentally undermined.

What societies demand

The last decade has seen enormous growth in the desire and need for electrical services (rather than generic 'energy'), and this has thrown environmental considerations out of the window across the world. But the pain is being felt most in developing nations, where funds are being redirected to electrical energy services; money which could otherwise be used for education, health and other necessities.

> The problems of the developing world are becoming worse – at a worryingly fast rate – because of the shortcomings of the grid format

The demand is being met by poor solutions. Grids are being built, physically expanded or increased in capacity to meet the new demands of millions. Even when hydropower or geothermal are employed in place of fossil fuels, they still use the central generation model, separating the user from the generation and discouraging energy efficiency. In the case of hydropower, the dependency is shifted to water, which in some places is scarcer than fossil fuel.

These grids will attract more people, and will fail. And they do fail; I don't know of any grids in the developing world that match the standards of a European country or a USA state (in terms of long-term confidence in the quality and dependability of the resource).

In off-grid regions, the newly found need for electricity has massively increased the use of batteries – either small disposable or larger lead-acid vehicle types. Batteries are environmentally treacherous. They contain chemicals and materials which poison land and water. They are ridiculously inefficient, often requiring over 30 times more energy to make than they will ever deliver. Most batteries reduce in capacity with elevated temperatures and with age; in these circumstances, they deliver far less than their ratings promise. In short, they are a damaging and expensive way to enable electrical services.

Diesel generators, another disproportionately expensive way of obtaining electricity, are increasingly being used to charge devices or power digital services. In many locations fuel dependency is worsening, not improving, simply because there is a need to generate electricity. Hundreds of millions of dollars are being wasted in each of these sectors by people, companies and governments who desperately need to be conserving spending. But the problems go further than the financial and environmental. The geographical limitations of the grid are increasingly influencing the structure of societies.

Grids lead to urbanisation

We have already seen how the grid attracts people from all around. In poor countries, the cost of most remote energy solutions has always restricted vital services to urban regions at the expense of rural areas. In the case of healthcare, inadequate rural facilities result in increased suffering and death; while the future of nations is being restricted by shortfalls in education.

The draw of rural people to urban centres – be it for employment, food, healthcare, education or other services – brings added problems of disease, poverty and social unrest. Electricity access alone does not cause urbanisation; but it certainly accelerates it, once hundreds of millions of people aspire to personal communications and online services. The monumental rise in mobile phone use has amplified, and also highlights this imbalance, between those predominantly urban areas with the grid and those mainly rural areas without it.

Nowadays, small towns, centred on main artery roads and utility electricity spurs, are guaranteed to have mobile phone charging stations as well as cold drinks, TV and, increasingly, internet access points. One 'domestic' electricity connection can be charging 40 phones per day while running TV, radio and internet simultaneously – with the volume turned up nice and loud of course. An additional outcome of this, as with any consumption, is the perceived social status of being on the grid. One feels prosperous and special when utility power is acquired.

Fear of change

All of this will be self-evident to some readers. It may seem very negative to describe the limitations of the utility grid format in the way you see above. But after 20 years working within the developing world energy sector, I cannot overstate just how difficult it is to have a balanced discussion about energy services provision. The utility grid format is pushed onto developing nations regardless of its drawbacks.

There are many layers of resistance to alternative energy solutions, and dogged promotion of the grid. At its heart is our widespread lack of understanding of energy and engineering; combined with our natural tendencies to resist change, a sense of familiarity or 'business as usual'. Meanwhile there are powerful interests behind both fuel and grid infrastructure supply. In fact, the mainstream consensus would have it that we lack any alternative candidates for providing the energy services that people need. We shall turn to this next.

Endnote

1. Source: https://data.worldbank.org (2016–2017 data).

CHAPTER 4

Accepting that the grid format cannot end energy poverty

Abstract

Discussions regarding energy provision too often start with, or slip into, comparisons with the central generation grid model. This is a problem. In this chapter we explore how to reframe the discussion by analysing our needs.

Energy on its own does nothing; we need to apply that energy to various tasks. It is these electrical services *that matter; and their operation needs to be unrestricted, uninterrupted and reliable. As we have read in previous chapters, no electricity grid is capable of providing this in the developing world.*

It would seem logical to develop any alternatives to the grid from the perspective of fuel avoidance. However to argue the negative aspects of burning fuel is, while scientifically unassailable, simply not powerful enough for either the industrialised or developing worlds.

We conclude that there needs to be a different approach to energy services provision, and that approach must be holistic and centred on people.

The 'alternatives' dilemma

Discussions regarding energy provision too often start with, or slip into, comparisons with the central generation grid model. This is a problem. The grid is presented in an unreasonably positive manner as the foundation or mainstream against which everything is compared. This has the natural effect of reducing the comparative credibility of any alternatives as just that – complementary 'alternatives' to the unquestioned consensus.

Combined with a typically low understanding of the many energy solutions that can be used, 'true' alternatives – solutions with compelling strengths and advantages to challenge and ultimately discredit this mainstream utility grid approach – are unfairly perceived as less important.

In order to provide quality energy services, we must start from first principles, by analysing our needs. But as we shall see, when we think of needs, we are influenced by our mental and cultural familiarity with the grid, and the lifestyles it has enabled.

http://dx.doi.org/10.3362/9781788530705.004

Energy or services?

There are many ways of analysing the energy needs of rural people and house-holds in developing nations. Typical of these are the following:

- Assessments of the number of lumens of LED light needed to replace kerosene lamps
- The electricity requirement for a twin stove to replace biomass cooking and water boiling
- The energy needed to heat a cold home; or to cool a warm one
- Energy for equipment such as computers for schoolchildren, mobile phone charging and refrigeration

These studies make assessments of electricity requirement against the cost of meeting a minimum level of provision. (Some make technical assumptions regarding a device's efficiency, but they frame the requirement in terms of electrical energy.) However, when it comes to powering industry, cities or regional infrastructure, the priority is not to be restrictive, whatever the cost. This commonly equates such needs with the provision of *unlimited* electricity.

Whether this is a conscious or subconscious conformity with the indus-trialised world is debatable. In industrialised nations, the grid provides effec-tively limitless electricity; it is the only electrical option that can.

If you want to power all the services of a modern-day city, you are going to need substantial capacity to do so. And if you want to attract and retain affluent citizens, you will be expected to power all of their modern household appliances.

What is significant here is the universal assumption of unlimited provision. This line of thinking is flawed, because it misses a subtle, but fundamental point. We have got it into our heads that our national need calls for unlimited energy. But energy on its own does nothing; we need to apply that energy to various tasks. It is these electrical *services* that matter; and their operation needs to be unrestricted, uninterrupted and reliable. As we have read, no grid is capable of providing this in the developing world.

***

Energy on its own does nothing. It is electrical services that matter

***

Appreciating the difference between unrestricted electricity and the need for unrestricted services is critical to future energy provision. The former tries to replicate the industrialised networked grid model in places where it cannot be implemented, or throws money at the grid system to enable consumers to waste most of its product. The latter – providing unrestricted *services* – is not only possible for the majority of people who need it, but promotes greater levels of reliability and personal value.

This is not an easy proposition to accept. The approach to energy services provision is very different from the way in which electricity is supplied. It is a cultural change; resisted by vested parties, but which we will have to make eventually. Services are akin to products rather than commodities. Meanwhile, many stakeholders and commentators have strong convictions that it cannot be achieved. Or perhaps they would rather it didn't happen.

The language we use

Assumptions and vested interests may promote the central generation format above other options, but we too play a supporting role, in our language and terminology. Convincing people of the grid's shortcomings, or that renewables are better value than popularly believed, requires a change in the way we describe these sources of energy.

The two most common questions from the general public, when enquiring about domestic solar panels are 'How much will it cost to power my home?' and 'how efficient are the panels?' We focus on the total cost for meeting energy needs and the operational efficiency in doing so. But we talk of fuel or electricity mostly in terms of the unit price: $1.50 per litre of petrol; $0.10 per kWh for mains electricity. Notice that we don't talk of efficiency or the broader, practical expense; as if it were somehow taboo or off-message. Our choice of language, and frame of reference, helps to maintain the status quo: in other words, our society does a natural job of keeping things the way they are.

Even if we were to master the terminology and metrics, we would struggle to comprehend just *how many* people in the developing world need energy services. We can talk about a particular demographic (say, kerosene users) and comprehend it in the context of a region or community; perhaps, at a stretch, a country. But we are simply incapable of relating to the hundreds of millions of people and the wide range of ugly challenges in their lives, let alone how to meet their needs. And so our imaginations return to the great big familiar solution already in place – the grid model.

Renewable energy and interconnected grids

The apparent scale of the need for energy services is pretty intimidating, underwritten as it is by the diversity of applications for its use, the multitudes of users and their differing environments. It would appear to support the view that the grid is the only solution to meet the requirements of so many people.

This is a major barrier to promoting sustainable technologies, especially given our preoccupation with the central generation model. Credible and unbiased commentators agree that renewables are an essential part of the world's energy future. But there is disagreement regarding their effectiveness and value. Hydropower needs water; wind generators need sufficiently consistent wind; geothermal is dependent on low elevation and tectonic good fortune; and of course, solar is dependent on sunlight. Some of these are hard prerequisites: hydro generation doesn't happen without water; and locations are very limited. But solar and wind are *gradient* in potential – they will work effectively with relatively poor resource and their value increases with the quality of that resource. The variable nature of renewable technologies is considered a barrier to widespread deployment, but as you may have guessed, I do not consider this the case for solar PV.

I contest that although the grid format meets the needs of millions and even tens of millions, it cannot meet the current or emerging needs of *billions*. As it goes, this was true even for industrialised nations. Grids in Europe and the USA were not built as singular entities to serve the tens of millions of people they do today: they were developed organically; built first for thousands, then hundreds of thousands. These systems were later interconnected. All the while technical and capacity upgrades were implemented and, in some cases, entire power stations were replaced. In the heart of London, for example, Battersea Power Station (now luxury apartments) and Bankside Power Station (now the Tate Modern gallery) stand as two great monuments to the systematic growth of the electricity network and the industry as a whole.

Although the grid format meets the needs of millions and even tens of millions, it cannot meet the current or emerging needs of billions

---

None of this is possible in the developing world. The landmasses are too vast, and the terrains too prohibitive, for interconnecting grids. Where in industrialised nations they provide redundancy and control over localised peak demands, their equivalents in the developing world would just connect big problems together – with heightened mutual vulnerability as the outcome. The isolated grids, meanwhile, have enough trouble coping with existing demands.

Knowledge and the policy makers

To those involved in the provision of services and energy solutions to the developing world, an unreliable grid and the need for alternatives to fossil fuels might seem wearisomely self-evident. But there is still a widespread lack of awareness concerning the challenges; the understanding, the conviction and the collective will towards appropriate solutions have not yet filtered through to those who define national energy and infrastructure policy.

In the interim, then, as an example, there are high-profile initiatives in many poor nations to reduce the unhealthy, polluting and expensive use of cookstoves and kerosene lamps. Yet, in the same countries, huge fuel dependency and pollution are often being implemented in the form of new generating capacity for mains electricity systems; the highest percentage of which are based on coal. Here, then, policies are helping individuals while ruining nations. Policy makers are simply shifting the problem out of sight and arguably making it worse – which is exactly what we have done in industrialised societies.

Even when new generating capacity takes the form of clean renewable energy, when connected to underperforming grids there is a dirty legacy and a harsh reality to the value of these technologies.

Our misunderstandings are basic. For example, there is continuous surprise from people who learn that renewable energy systems connected to utility grids actually require the grid to be active. In short, if for any reason there is a power cut on the network, the renewable energy systems will not act as a backup device. If a domestic home has solar panels on the roof generating electricity, the potential power cannot be used within the property if the utility electricity fails. This is why domestic energy storage systems are growing so rapidly in popularity. However, these are for short-term resilience against grid failures of a few hours, or as a means of reducing costs.

Grid-connected renewable generation (such as wind, solar and micro-hydro) not only require the grid to be present in order to supply power into it, they also require the quality of the grid – the voltage and frequency – to be within pre-defined limits, or the renewable generation, by law, must automatically disconnect itself.

The challenges of distribution

Policy makers are slowly accepting that spending large amounts of money on the central generation format is futile; even when it is served by renewables. But the more recent industrialised world approach of strengthening existing grids with distributed generation is still viewed as a viable solution. As before, the basic concept makes sense, but the devil is in the detail.

Complex grids are kept operational using sophisticated control systems. These balance multiple sources (i.e. power stations) to serve millions of consumers on a second-by-second basis. The basic arrangement is to use 'base-load' generation to meet the majority of demand all the time, while employing other sources of generation to meet 'peak', or variable, demand. This variable demand can be anything from a few per cent to sometimes over 30% of the total. A nuclear power station, for example, will generally be used for base-load demand, whereas a gas-fuelled power station (which can increase or decrease output relatively quickly) can be used for both base-load and variable demand. Large hydro can also be controlled relatively easily and used to address imbalances in the grid. Base-load generation ensures that distributed generation remains operational. The grid is kept within tight operational parameters and very rarely fails.

Any distributed generation (such as domestic solar) can feed their generated energy into the grid network for general consumption. Meanwhile, larger wind farms and solar PV arrays can be used to help stabilise grid quality; but they must be sufficiently large relative to the local demand and represent a meaningful percentage of the regional generating capacity. However, in the developing world, distributed renewable energy generation is far less valuable. These systems feed into a grid whose voltage and frequency are unstable. They therefore cut in and out of operation as the quality of the grid varies. This is a major issue for grid-connecting small solar power systems and wind turbines. (The alternative is to widen the operational parameters of the distributed generators, but this allows poor quality mains to damage the generator control systems, which in turn reduces their efficiency and operational lives.)

Large distributed generation *can* help to stabilise the grid, and it is being installed across the developing world to do so; but this is a short-term solution. When the base-load grid fails, or is pulled out of its operational parameters (by excessive loading), this causes a domino effect and the distributed generation shuts down in response. Trying to get the whole grid back up is by no means straightforward.

This problem can largely be overcome, by adding energy storage to the mix; almost always in the form of batteries. When the quality of the grid goes out of tolerance, the battery is able to supply power to the property or user. This is an 'off-grid system' with a mains utility connection.

Energy storage is valuable; but there are practical and commercial limits to their capacity and subsequent ability to stabilise utility spurs or larger grids. This is another extremely expensive and temporary fix to an inappropriate energy source.

Any renewable generation technology, connected to
the grid without storage, is subservient to dirty fuel

Investing in fuel dependency

All this aside, ploughing millions into grid-connected renewable generation in the developing world is wrong for a very simple reason: it makes fossil fuels essential. A high percentage of base-load generation on developing nation grids is provided by fossil fuelled power stations, mostly fuelled by coal, our dirtiest energy source. Any renewable generation technology, connected to the grid without storage, is subservient to this dirty fuel.

After all the rhetoric of reducing pollution and overseas dependency, the very initiatives that some think are helping the poorest of our world's nations reduce their overseas fuel dependency are actually guaranteeing that dependency into the future. A cynic might think that fuel suppliers are facilitating limited amounts of renewable energy for this very reason.

The classic old banger

If you are rich and have a classic car, you can re-live a romantic attachment to the past and enjoy the distinctive and contrasting experiences it allows. If you have a 30-year-old car simply because you can't afford to replace it and need it for your working life, it is a continuous challenge just to keep it functioning.

Ploughing money and effort into grid systems is a necessity across the globe, just to keep them working. But trying to expand them or make them more reliable is ultimately a fruitless exercise, particularly in the developing world.

When it comes to utility supplies, industrialised nations will seemingly carry on wasting huge amounts of money by being reactive rather than proactive.

This is despite the spiralling financial costs and the increasing vulnerability to system failures and fuel supply restrictions. The climate, meanwhile, doesn't even seem to feature as a notable motivation factor anymore.

Meanwhile, the developing world does not have this money to burn, and doesn't receive sufficient value from the grid. More than any of us, it needs to be released from dependency. We need to redirect the enormous amounts of money being thrown at grid systems. Politically this is neither expedient nor attractive, because funding for *reliability* of electrical supplies doesn't carry the *appearance* of progress, and therefore doesn't stimulate response. 'Connecting a million rural farmers' or 'providing 25% of schools with electricity' are the common straplines for finance negotiations. Once physically connected, funding attention moves to another area, regardless of how little value has actually been provided.

Changing our emphasis from this reactionary expansion of overburdened energy supplies, to a forward-thinking proactive programme of service provision, represents a Herculean challenge; and it is this to which we shall now turn our attention.

Thinking about being proactive?

It is time to overcome our fixation with utility electricity and central generation. Time to question the prevailing culture, and stop distracting attention from high quality solutions. In short, it is time to stop building more grids, or patching up existing ones in vain. Instead of spending tens of millions of dollars on central generation or even large rural mini-grids, we must use that money and direct our efforts to the following ends:

- We can bring generation and consumption, in every sense, *closer to the user*. Closer in proximity to promote self-determination and independence; and closer to the lives of everyday people, fostering a true appreciation of the energy we use.
- We can turn our efforts to social change: in promoting awareness and efficiency; particularly *fuel* efficiency – which increases the value of energy services.
- We have the opportunity to remove the need for fuel altogether. Energy solutions which tap the free, clean, infinite resources around us, and which herald an escape from a dependent, polluting, energy poverty trap.

Addressing the grid

What can we do about existing grid systems which are unable to cope, either now or in the future? How do we deal with the increasing numbers of people being drawn to the grid? And what is the best way to address existing energy services in the developing world?

Remember that we have exposed the grid for what it is; more than a mere protocol for transferring energy services, it is a social construct, a way of thinking; one which is hostile to alternatives; and yet we have much higher value options to hand. They are not waiting to be invented; only for us to establish them adequately in the real world.

Our first step, then, is to make these non-grid energy solutions widely available. The core aim is to reduce grid consumption by second-tier consumers – those who access someone else's grid connection (either at a cost, with their consent or illegally). The concurrent aim is to stop more people connecting to grid electricity.

Secondly, broaden the reach of those higher value solutions to reduce the consumption of existing grid users and, ultimately, take people off the grid. This will enable the existing generation and transmission infrastructure to work in the interests of dependability, rather than mere numbers of poorly served people.

Solutions among the four categories

Consider again the four distinct electricity categories we defined in Chapter 1: reliable networked grids; unreliable networked grids; limited mains access; and off-grid (those with no reasonable access to mains electricity). Without too much of a stretch, we can see that the two categories in the middle share a common ground, at least in terms of practical capability to use electrical services (that is, if we talk about quality and dependability). People who presently have *unreliable networked grids* need the same solutions as those with *limited mains electricity access*.

At some point in the future, the solutions designed for these two groups will also become attractive to those who presently enjoy *reliable networked grids*, the majority of whom are in the industrialised world. The time will arrive when we release ourselves from grid and fuel dependency; be it by choice or necessity. This, therefore, is a long game. Where does it start?

The time will arrive when we release ourselves from
grid and fuel dependency; be it by choice or necessity

The off-grid developing world

In the developing world, everyone else outside of the first three categories is off-grid (with no reasonable access to mains electricity). Autonomous energy solutions are currently targeted at rural off-grid regions. However, I do not believe that this is the most effective way to propagate energy services.

The scale and diversity of second-tier grid users may be daunting, but it is a walk in the park compared with trying to reach all those who are off-grid. The problem with off-grid populations is that it is so time consuming, expensive and logistically challenging to reach meaningful numbers.

Consumables and products typically find their way to off-grid regions by starting in urban centres; only later filtering into the formal and informal national distribution channels. This is true around the world: consumer goods are first made available in cities and the highest population density regions where there is the strongest revenue potential. As supply volumes increase and local distribution networks respond to demand, products propagate outwards; first to semi-urban, then to rural centres and finally to small, widely dispersed rural outlets.

The catch is this. By definition, the value of off-grid products is highest the *further away* you are from the mains electricity system (which, as we have noted, has a strong urban bias). So product solutions designed specifically for off-grid regions often have little value in the urban areas in which they first need to be sold. This has prevented many a good product from becoming established in the off-grid areas where they are most valuable and most needed. A narrow focus on off-grid needs, then, will always be met by natural barriers of spending power and excessive cost of sales due to the geographical and cultural diversity of its people.

Instead, I believe that we need to utilise the spending power of the more affluent sector, in order to bring *quality* products (and the networks to supply and support them) into poor regions. In doing so, we make access to energy service solutions easier and more valuable for those who need them the most.

This, though, is not to ignore off-grid people in the assessment of requirements. On the contrary: they are our most vivid representation of what the world will be like when we no longer have widespread utility grids. In many ways, the off-grid peoples of the world help define the needs for a fuel-independent future.

We need to utilise the spending power of the more affluent sector in order to bring quality products into poor regions

---

Meanwhile, controversially, the great 'advantage' of the present energy-poor of the world is that their needs are very modest. Meeting them with robust and scalable stand-alone solutions proves far easier and cheaper than trying to match the excessive energy demands of those presently enjoying reliable networked grids, and provides a solid basis for progress.

Learning from the grid

We have explored the shortcomings of the grid format. The rationale for entering into such a lengthy analysis of the grid over the previous pages is to turn it to our advantage. What we find thoroughly discredits its claim to unchallenged dominance. Energy service solutions should be defined by needs and wants – as valid in their own right and not as 'alternatives to the grid'; but it is inevitable that comparisons will be made. In doing so, hopefully this section will act as a guide, both to understanding the logical arguments which refute the notion of all-conquering grids; and to the unsung approaches to energy services which are actually more likely to succeed.

The issue of fuel

In the developing world, the most commonly cited limitation of utility electricity is its restricted geographical reach; hence the common distinction of off-grid and grid-connected communities. Any discussion regarding off-grid regions will need to address energy, which as we know is popularly synonymous with fuels. In grid-connected regions, the issue of fuel is not so prominent; nevertheless, it is among the primary environmental criticisms of utility grid systems, along with national dependencies on unfavourable foreign suppliers of fuels. (While cost is a concern, it is not sufficiently potent to prompt action or meaningful change; at least, not yet.)

Fuel, then, is the common bad guy for rich and poor nations, for off-grid and on-grid people. It would seem logical to develop any alternatives to the grid from the perspective of fuel avoidance. But I consider this a mistake. Of course, reducing the use of fossil fuels – the pollution they create, the environmental, financial and overseas dependency issues they bring – is critical to our future. But the issue of fuel has not proven a strong enough motivator for change. To argue the negative aspects of burning fuel (or the benefits of offsetting its use) is, while scientifically unassailable, simply not powerful enough.

This has been proven time and time again in off-grid regions. I'm not just talking of the questionable benefits of wind-up radios or solar cookers; I mean convincing and lasting solutions such as micro-hydro and small wind turbines to replace diesel generators; or, more significantly, products such as solar LED lights to replace kerosene lamps. After over a decade of these solar products being available there likely remains over $20 billion spent annually on kerosene for lighting[1] – mostly by poor people.

Yet, for these products, *the financial value propositions alone* are incredibly high. For example, a good solar LED light can cost the equivalent of 3 months' worth of kerosene, and can last for 5 years with zero running costs. All of this is before you even consider how safe, clean, rugged, portable and versatile they are compared with the naked flame lamp they replace.

What about proactive engagement in the international community? When I first became involved in the clean energy movement in 1999, we treated the avoidance of polluting fuels as an inspiring opportunity to prevent climate change – confident as we were of the tipping points and positive feedback mechanisms in the data to hand. But the language of climate change shifted; from prevention, to deferral, to mitigation, to now shrugging and watching it happen. Political justifications for reducing fuel use on environmental grounds are now so long-standing that a once-logical discussion is now characterised by reactive measures; every bit as ineffective as the cost arguments in driving strategy, action and change.

Within international politics, the principles of clean energy have been hijacked by vested interest groups. Two examples stand out in my mind.

The myth of clean fuel

The first is the myth of clean fuel. There is no such thing as 'clean coal'; it is just a cleaner form of the dirtiest fuel we have. The raw fuel for nuclear power is mined at great environmental cost and leaves a terrifying legacy for the future. (I have never understood how these two discussions ever gained traction, but maybe that is the fault of people like me, who simply considered these, and many other proposals for abating climate change, so ridiculous that no one would ever take them seriously. Yet here they are, in the mainstream of the debate.)

Neither fossil fuels nor nuclear can be clean or sustainable, period. If you are someone who feels passionately that this is not the case, please stop reading; this text is not for you. Proposing clean coal and nuclear as solutions to global pollution, along with climate engineering and space-based energy harvesting are best put into a single group. I suggest the title of this group is 'crazy and ridiculously expensive initiatives that make our problems worse'. Feel free to think up your own.

Overseas dependencies

Meanwhile, the non-environmental justifications for avoiding fuel dependency should be sufficiently self-evident to effect change. But the motivation for change has proven weak in the face of international interests. The industrialised world has actually *increased* its consumption of fossil fuels from non-democratic nations over the last three decades. Bear in mind this increase continued through the period when the implications of dependency have been most gruesomely demonstrated in the USA and Europe. In 2014, Russia's annexation of the Ukraine provoked little in the way of response from the international community. Why? The fact that Russia supplies around one-third of Europe's gas will provide a clue, but reasons aside, the momentum of overseas dependency combined with political intransigence proves a powerful force.

We are all dependent

Of course, no country produces everything that it needs anymore. As the modern world becomes increasingly interconnected, we have become reliant on overseas suppliers for numerous aspects of our lives. Many would like to avoid having to provide huge sums of money and political power to countries whose policies differ from our own; but the way we have structured our global society makes this unrealistic.

The key is to utilise what is available in the world, without becoming dependent in the long term. This means not relying on resources that are only available from a limited number of suppliers, particularly if you need a large amount of that resource. This is tricky if you are projecting your aspirations onto a distant region. If a person is drowning, you throw them a life saver; you do not withhold it while espousing self-sufficiency. Similarly, we cannot in conscience preach the risks of dependency to less powerful nations, punishing them for disrupting the global climate, or our own vested trade interests. On the vexed question of energy services for developing nations, the only initial option is to source key components internationally. But in order to reduce dependency, and eventually remove it, we have to be as effective as possible with the sourced goods. This means that you have to bring the skills required to develop them in-country.

The key is to utilise what is available in the world,
without becoming dependent in the long term

Personalising energy services

As we know, those of us with reliable networked grids are emotionally and physically disconnected from the fuel that is used to produce our electricity. Expending no effort in obtaining electricity, we have no understanding of its complexity or how it is delivered to us. Assuming an unlimited supply, we place no value on it. It needn't be this way. We can address these issues when we attract people away from grid electricity, in the very energy services we develop.

Energy literacy is a necessity. It should take a prominent role in any strategy we care to bring to the table. We know that converting knowledge into pro-active consumer change is a frustratingly slow process; so it is going to take a bolder, more assertive approach to convince people to value energy services.

Far from chasing a mythical 'alternative' to replace the ailing central utility grids of our own experience, there exists a more comprehensive, holistic approach to engaging people in the means of their own energy consumption. It concerns the following:

- The way a product works out of the box
- The way a product is administered by an individual or community
- The way a brand is marketed to a new sector
- The methods of maintaining the services, appliances and devices

These are all areas which, rather than being fed to a passive, compliant customer, instead derive their marketable strength from the *level of engagement brought by that user*. The user must understand potential value pre-sales, and will only receive that value by understanding how to operate the energy service. Commercial success is dependent on this engagement.

Going off-grid, we are bringing self-actualised capability into the mix; and with it, an intrinsic *educational* requirement to deploying such systems. The principles of brand value and brand loyalty can be brought to bear on this educational requirement. Looked at another way, the marketing strategy itself can be educational in essence.

We must therefore put the user at the heart of any solution and encourage a self-determined level of appreciation of the resource they have. We need to *personalise* energy services.

Endnote

1. Derived from multiple articles including: https://www.sciencedirect.com/science/article/abs/pii/S0305750X17302103?via%3Dihub and https://www.climatechangenews.com/2013/02/20/un-launches-new-africa-plan-to-swop-kerosene-for-solar/

PART TWO

Introducing the Selling Daylight proposition

CHAPTER 5

Personalising energy

Abstract

The tenets of 'personalised energy' demonstrate our plan to reach hundreds of millions of people with dependable energy services, the fruits of which can become as ubiquitous as the mobile phone, the newspaper, the bicycle or the water pump.

In this chapter we look into the top-level requirements, or core values of personalised energy:

- *Involving the user in both generation and consumption;*
- *Promoting value: encouraging the user to value the electricity they have at their disposal;*
- *Sharing knowledge and understanding of energy services.*

We explain why limiting supply for the user does not have to mean limiting the service that energy is enabling.

We introduce the need for user interfaces, intuitive hardware design and most importantly, a way to convey information to absolutely anyone; neutral in terms of culture, religion, ethnicity, gender and age; as well as language, literacy and aptitude with technology.

For many years I have studied the way people use energy, and how it benefits their lives. I have come to recognise that, collectively, we are too demanding and wasteful. For those of us lucky enough to have reliable utility electricity, the financial, environmental and human costs of using a centralised system are simply too high, as we become globally dependent on electrical services, all the while pushing the limits of our present infrastructure.

I have come to believe the following: *personalising energy is essential.*

Consider any service that requires energy. (Your example could serve a traditional domestic need, such as cooking; an emerging application, such as transport; or some hypothetical future use.) This service will contain, somewhere, a maximum point of value: a perfect expression of its efficiency, cost, quality and convenience; in short, the greatest benefit it can deliver.

This optimal value is real, not fictional. And for a given product, it can only be understood and attained by the person who uses it. A product's lifespan, versatility, performance and value for money are all inextricably linked with its user's engagement, appreciation and involvement. If we can ensure sufficient understanding of the necessary aspects of energy services, while providing the tools for controlling them, we can optimise both the cost and the value of the

http://dx.doi.org/10.3362/9781788530705.005

solution. Putting users at the centre of their energy services – both physically and emotionally – makes that task easier to envisage in the absence of a dependable utility grid. If we endeavour to engage the individual in this process, we open the door to many opportunities and rich rewards. With it, we reveal new challenges.

The section which follows is an attempt to define 'Personalised Energy' as an initiative, akin to a brand. This is a set of guiding principles which exist at the very heart of the concept of 'Selling Daylight', to the point where the two phrases are almost interchangeable. It is a statement of intent which positions the gigantic demographic of end users at the heart of our solution.

The tenets of Personalised Energy demonstrate our plan to reach hundreds of millions of people with dependable energy services, the fruits of which can become as ubiquitous as the mobile phone, the newspaper, the bicycle or the water pump. It is a phrase which we aim to place firmly into the cultures of the world, so that it becomes self-evident, synonymous with energy itself, supplanting its traditional connotation with fossil fuels. If users are to take centre-stage in any energy solution, they will need many aspects of system functionality and value communicated to them. The better we convey this information, the more value our solutions will provide; and the better they will be considered by others.

The true value of a service is never fully appreciated unless there is some sort of effort involved in obtaining that service, or unless there are direct personal consequences of not valuing it. This effectively means that we face imposing realistic limitations on the user and bestowing a degree of personal responsibility. I will make the case that this is a positive thing.

Put simply, the goal of personalising energy raises the following basic imperatives. We can treat these as the top-level requirements, or core values.

- Involving the user in both generation and consumption;
- Promoting value: encouraging the user to value the electricity they have at their disposal;
- Sharing knowledge and understanding of energy services.

The phrase 'Personalised Energy', like any high concept, carries with it a latent contradiction. I have already said it is the reliability of 'services', rather than 'energy', which matters most. But I believe that when we talk of 'Personalised Energy' we are addressing the user as a central player unlocking the understanding necessary to draw tangible, meaningful benefit from energy, in the form of dependable and reliable services. While we're at it, the name 'Personalised Energy' also relates to where funding for electrical services will inevitably come from: energy budgets.

Involving the user in generation and consumption

With a view to personalisation, energy should be generated as close to the user as possible. This does not have to be literally right in front of their faces, nor necessarily about their property; but it should enable some sort of visual appreciation at the very least, to encourage people to learn about these systems. This already works for people who live next to large power stations, or have electricity pylons in their back yards. The intimidating proximity of the infrastructure does wonders in this regard.

Such proximity enjoys a direct relationship with scale. Generation does not *have* to be on a small scale in order to be appreciated; but it generally helps, for the simple reason that the user will likely have a far more personal relationship with their device. A good analogy is the relationship we have with the energy consumption of our smartphones, compared with that of our desktop PCs. Smaller energy solutions generally yield greater levels of appreciation – the closer a user gets to the workings of the system, the greater their awareness of its operation; and, importantly, its efficiency.

Generation and daylight

If we are generating electricity near to the user, we need a resource which is versatile and accessible. This gives us two options. We could use combustible fuel, which allows us (in theory at least) to generate electricity at any time. Alternatively, we can use *variable* energy resources combined with an energy storage system. In a global context, 'variable' resources mean wind, water or sunlight. 'Storage', again for our widespread deployment needs, essentially means batteries.

Sunlight is the only sustainable energy resource that can be used almost anywhere. Daylight, after all, is very predictable, and the technology for harnessing it has been in widespread use for over 30 years. This is the primary reason for positioning solar photovoltaics at the centre of our strategy.

The user interface

Involving the user in consumption can be very simple. Any battery powered device (such as a mobile phone, laptop or music player) demonstrates this; its charge indicator or icon showing us how much remaining energy we have. We adjust our habits accordingly with a view to when we can next charge that device.

Such indicators fall within the broad discipline of *user interfaces* (UIs). These represent the meeting place between humans and technology; and they are central to the way we promote understanding and engagement in users. For our purposes, talking of display or feedback mechanisms, a user interface can take many forms. It might be a large display on a wall, a smartphone app, numbers on parallel wheels (old domestic utility meters) or a simple set of LED lights of different colours.

These interfaces are available for utility electricity, water, gas and other consumption metering, but by far the most valuable application is for battery systems; simply because they have a limited supply of energy, and the UI enables a degree of awareness, and therefore limited control, of that supply.

A user interface has to be universally accessible. Realistically we cannot educate each user face-to-face about the technical operation of their electrical systems, so we need them to understand generation and consumption at the point of use, instinctively and with the minimum cognitive effort.

User interfaces are central to the way we promote understanding and engagement in users

---

Image-based UIs provide the universal answer we need. To gain the most from our portable consumer electronics we don't need to learn about ampere-hours and the technical issues affecting battery capacity; we just read a little icon on the screen. The same principle can apply to our domestic energy solutions.

But user interfaces do more than merely inform us of immediate events; they also help us to learn over time. A smartphone, for example, informs you visually that if you stream video from the internet it will drain the battery faster than just sending text messages. By showing a rate of change, the UI – the charge indicator – is a simplistic but effective consumption regulator which leads to a change in baseline user practices.

Promoting the value of energy services

What is the true value of an energy service? Or any service at all? If we don't fully understand the commodity we are buying, or have not participated in any meaningful event in order to acquire it, all we have to go on is a gut feeling, an incomplete emotional response; perhaps, an inadequate tick box (zero, not satisfied; to five, extremely satisfied) on a customer questionnaire.

The word 'journey' is widely overused in modern popular culture, often describing a TV show contestant's emotional rite of passage to tearful victory, or equally tearful exclusion. But it conceals a truth of sorts; that we make sense of our world, and thereby derive a sense of value, by way of narratives and stories, however small. This holds true for everyday actions and transactions, including the way we approach essential commodities. Energy services are no exception. In order to appreciate its value, the user needs to have experienced some sort of meaningful event in obtaining that service.

Simple cause and effect, an action and its outcome, read back to you – reveals its significance; and of course, the experience involves effort, which reinforces it. Conversely, the user may ignore, or fail to value the benefits of an energy service; and discover the consequences at a direct personal cost. Either way, there must be a visible and tangible impact to the user. If our energy services are to be attractive to consumers, we must not hide this cost; but we must not allow its impact to be too harsh. Originally, I had assumed that this would be quite a straightforward affair. On the face of it, simple tasks, such as taking the solar device outside each morning for charging, or cleaning its panels on a weekly basis, would appear to suffice. But as it turns out, there are several levels of complication to consider; when we talk of 'effort' we mean more than just physical demands.

Consequences: cost

The obvious consequence of using an energy service is financial: if you waste lots of energy, you pay for it. But as we know, this has not been an incentive for those of us with reliable networked grid electricity. And despite the costs, our practices in general show that money is not a strong enough incentive. Consider other wasteful practices in the face of visible consequences: for instance, how much food we throw away; or how many consumer electronic devices we purchase. Consider also everyday events, such as sitting in a stationary car for extended periods of time with the engine running. These are all too human traits and they manifest in different ways with different people. I have learned how these human traits are exhibited in the form of illogical practices by those new to grid electricity.

Irrational consumption

You might think that, in principle, if someone with limited financial means chooses to pay directly for the energy they use, they are likely to be more efficient with it. But this is not always the case. When a household has electricity for the first time it can be used as a status symbol, especially if not many others in the area have it.

I can remember the novelty of buying my first camera and taking photographs of everything I saw, even though it used film that had to be processed at quite a cost. The same is true of those older than me who had to use single-use flash bulbs. Nowadays of course, digital pictures do not cost anything. But my point is that when people have electricity (particularly relevant if we consider that it is generated using an old-fashioned and costly technology), they can use it beyond their means and consume themselves into debt. Whether they leave the lighting on all night (for security reasons, or to show their neighbours that they have electricity); or whether they leave the TV on all the time; let everyone come in and charge their phones; or leave the fridge open for long periods; it all adds up.

People who are acutely aware of the value of raw fuel may, on the other hand, fail to understand why, say, the door of the fridge needs to be kept closed – or the cumulative cost of leaving lights on at night. (This, incidentally, pretty much mirrors what happens in the comfort of affluent countries. When someone opens the fridge door then turns and starts a conversation I basically start hyperventilating as I can almost *see* the energy falling out.)

People who are acutely aware of the value of raw fuel may not understand why the door of the fridge needs to be kept closed

---

Metering and monitoring

Companies who sell electricity do not make it easy to keep an eye on how much of their product we are consuming. This will come as little surprise. From an interface perspective, electricity meters seldom assist us. How should we address the mains consumption issue in a way which draws our attention and promotes our involvement? A basic suggestion is to put the meter in a very prominent position. Make it big and illuminate it. Don't employ some little device outside the house or under the stairs (familiar to industrialised countries) for which you have to strain to see the numbers, let alone interpret them. Furthermore, we could make the meter a bar graph, and a prominent feature.

Rationing grid electricity

It is appropriate at this stage to make a bold point; *limiting supply for the user is not necessarily a bad thing.* Though it will strike some as heresy, the idea of a finite supply instantly communicates value to its consumer; the value of a commodity which requires intervention and management. Contrast this with what we have shown of centralised grids; invisible, inscrutable and inviting only complacency.

Besides which, rationing the grid supply would drastically improve energy efficiency in terms of both equipment consumption and user practices. Some energy specialists predict that we could save up to 50% of all grid electricity consumption through equipment and user efficiency. (Considering the inefficiencies of grid systems, that is a huge fuel saving.)

Grid electricity rationing is already practised regularly around the globe, and can be highly restrictive to users. This is far from an ideal scenario. However, not all rationing has to be inherently prohibitive. Outside of rich societies, for example, mobile phone use has operated under a form of rationing from the start. Talk time for mobile phones and, more recently, data bandwidth for smartphones are nearly always pre-paid, and devices often feature less battery life than those of industrialised nations (for reasons I will explain later). In addition, charging the device may prove difficult, time consuming or expensive; some people will keep the phone off most of the time to save battery life, switching on occasionally to send and receive texts (I know people who can make talk-time and data credit last for weeks this way). Charging might mean walking an hour to a charging 'facility' and then hanging around for several hours.

In such cases, having a quota of energy or user time might mean the user is efficient with the device, but at significant inconvenience. A balance is needed, therefore; solutions which avoid restrictive effort, time and cost, but which do so in a constant dialogue of supply and value with their users. Alas, for political reasons alone, it is unlikely that any form of practical rationing will happen in industrialised nations, unless no other option presents itself.

Limiting supply: the use of batteries

When people only have a limited amount of something, they are naturally more careful in how that 'something' is used. One of the best ways of drawing users' attention to consumption – thereafter to energy and financial efficiency – is to give people limited amounts. The battery is the defining component for our solutions. It meets the basic premise of ensuring that users invest something of themselves in energy services, and thereby derive value.

There are challenges along the way. For example, a very small electrical load (such as charging a few mobile phones) and a very large battery (from a truck, say) may not prompt the necessary level of personal involvement for each phone user, and arguably may fail to provide the incentive to value the energy.

But viewed more positively, we can see that *limiting the energy resource does not have to mean limiting the service that energy is enabling*. The trick is to have the user ensure they make the most of the limited energy at their disposal; so the first requirement is to ensure the size of the battery is carefully tailored to the services being powered. There is an inbuilt limitation to the service, which naturally encourages efficiency of consumption. But we can avoid unrealistic expectations, which would discourage user care. Put simply, the user must manage consumption to heighten the value of the service but must not be penalised too severely if they wane.

Meanwhile, the size of the battery also dictates the size of the PV to charge it, for a given location. In fact batteries define many of the requirements for our solution, such as cost, physical size, versatility, operational life and ease of use.

The battery is the defining component for our solutions. It meets
the basic premise of ensuring that users invest something of
themselves in energy services, and thereby derive value

Sharing knowledge

The third and final element in our quest for 'Personalised Energy' is knowledge. How do you go about ensuring that people will understand the difference between 'energy' and the services it provides? Unlike our two previous requirements, there is less received wisdom on which we can draw. In the discipline of knowledge sharing, there is no established method as such, and there is limited precedence from which we can learn and adapt.

Much can be achieved by smart design. Ergonomically and visually, our products can inform the user of what the system is doing, and can help ensure that daily operation is as intuitive as possible. But this alone cannot teach how to install a system, or how to obtain the best and longest life from it. Deeper challenges, such as how to reinforce abstract ideas of value, may be beyond its remit.

Those we aim to reach number in the hundreds of millions, divided by language, geography, culture and literacy. For our technical solution, our chosen energy resource is daylight, because it is universally accessible. Similarly, in determining a means to educate, inform and share knowledge, and without the means of language, we must find or develop an equivalent, where information enjoys a universal, common baseline. I call this a *Visual Instruction System* (VIS), and it should apply to all aspects of our products' non-verbal link to the outside world.

I explore the detailed challenges of visual instruction in a later chapter. For now, we can grasp that its central purpose is *universal reach*. We must be able to convey information to absolutely anyone; neutral in terms of culture, religion, ethnicity, gender and age; as well as language, literacy and aptitude with technology. To this end, it must be suitable for deployment over diverse media, from the traditional two-dimensional, static, paper-based drawing to the cutting-edge digital animations, storyboards and videos familiar to the contemporary digital world. Finally, we must be *engaging*; we need people to *want* to learn, otherwise we will never truly connect with them.

A uniform, consistent, visual communication system is, if it functions correctly, much like our chosen energy source. It is universal; it is ubiquitous; and is the base resource from which we can generate awareness. 'Knowledge sharing', then, asks more of us than mere tuition, a paper sheet stapled to a delivery box, or a 'Frequently Asked Questions' page. It should be a core brand value in its own right, and form part of the backbone of an energy strategy.

Personalising hardware and knowledge

In the broadest terms, energy services require both a hardware element – the physical equipment, including the logistics of how it reaches the user – and a 'soft element', meaning knowledge. I am proposing stand-alone PV systems for our hardware demands, and a universal visual media strategy to share knowledge. The challenge before us is to develop both of these in a way which makes them personal, valuable and relevant to anyone who uses them. Our target audience knows it needs reliable energy services; but does not yet know that solar PV is the credible option.

Is this truly possible? Can we provide energy service solutions for this many people? And can we give users the intuitive means to get the optimum value from our products? At the same time, are we equipped to convey operational principles to our users; or to communicate compelling financial value arguments to suppliers? I have discussed these universal communication needs with several language teachers and media artists. Unsurprisingly, all have stated that it is not possible to convey *detailed* technical concepts to such a diverse group.

But this is not strictly what we are trying to achieve. When a product is designed well, and is intuitive to use, only an essential minimum of information is required to ensure effective use. The rest of this book details how to achieve this and spread the concept to everyone.

What has surprised me most while looking at these needs is that, while getting the message across is a considerable challenge, the hardware is actually quite achievable. The technology exists and can be deployed with existing energy services expenditure.

Hardware: the consumer format

It is unreasonable to expect everyone – rich or poor, educated or not – to read instruction manuals or learn technical principles. As far as is possible, the hardware, and its operating system, must shoulder some of the burden. It needs to be engaging, intuitive to understand and simple to use. It is first imperative to look at what the hardware is capable of achieving, before filling in the gaps with communication and learning materials. Essentially, we need to tailor the hardware to meet the limitations of communication, as opposed to the other way around. All this is a long way of saying that *user-centred design* is essential, not a commercial burden.

With this in mind, the easiest way to think of the requirements of the hardware is that it needs to be in a *consumer product* format for the general public or in a *commercial solution* format for companies. That is to say, there are no skills required of the user that they do not already possess or can easily learn. For a typical person or household, these products should be no more complex to operate than a TV, media player or smartphone. For a typical commercial organisation, imagine a telecoms company, or a supplier of large refrigerators. These types of organisations will be entirely capable of assembling electrical components and carrying out such work as mechanical installations and commissioning. Our intended users should not need to have detailed knowledge in such technologies as batteries or solar PV.

These products should be no more complex to
operate than a TV, media player or smartphone

Autonomous solar photovoltaic systems are, essentially, automatic battery chargers. PV panels generate electricity when exposed to light and that electricity is used to charge a battery. This then supplies energy to the user device or equipment.

I believe stand-alone solar PV products and systems should be at the centre of energy service solutions for the developing world. In broad terms, they are the only format of energy which fits the requirements and works within the restrictions I have outlined.

This is not about being green, or representing higher financial value (although they are, and they do). The choice of stand-alone PV is about putting the user at the centre of the solution. They are compatible with the consumer product format and can therefore be manufactured, distributed and operated as such. The technical demands of PV systems do not trouble us unduly. The problems lie with perception – their reputation, and how they are presented to the general public and policy makers alike.

Before describing how we tailor stand-alone PV systems to our needs, let us first take something of a detour, to clarify, as a generic technology, exactly what solar PV is, and what it is used for.

CHAPTER 6

An introduction to PV systems

Abstract

This chapter describes what stand-alone PV systems are, their core value and the many ways in which they can be utilised.

Also known as 'off-grid' these systems are, in basic terms, automatic battery chargers. They operate autonomously, with no moving parts.

It is the comparative cost of the alternative *energy solution that defines the value of a stand-alone PV system; if there is one. The challenge is to represent and convey these savings adequately and comprehensibly.*

Stand-alone PV represents the greatest commercial opportunity within the developing world energy sector.

We describe grid-connected PV systems in moderate detail, primarily to emphasise the important differences from being 'off-grid', but also to highlight how the value of PV technology is being undermined.

There are more applications of stand-alone PV systems than there are apps for smartphones. Included is a categorised overview to convey an appreciation of the scope of what a versatile and autonomous energy solution can enable.

Stand-alone PV systems are often simply referred to as 'solar systems' or 'solar products'. 'Solar' is one of those over-used and abused words: I have seen it written, for no apparent reason, on pedal bikes, hosepipes and air fresheners. It means 'relating to or determined by the sun', and can describe such subjects as our planetary solar system, the solar calendar, or solar energy; and, of course, solar technology. The last of these contains a great deal of variety. But by far the most common examples are *thermal* and *photovoltaic*s.

Solar thermal technology utilises the Sun's energy for heating and cooking. Photovoltaics, or PV, is the direct generation of electricity from light. The word 'solar' is commonly used to describe both of these sectors, so it's worth mentioning this distinction at the outset.

Photovoltaic technology operates on a principle known as the *photoelectric effect*, and is a function of the materials used. When exposed to light, PV materials generate electrical charge. If connected within an appropriate electrical circuit, such as to a battery, then electricity will flow. The smallest PV component is a PV *cell*, which generates a relatively low voltage and current. PV *modules* (often called panels) incorporate a number of cells to form higher, more useful electrical capacities. They come in numerous shapes and sizes, but as far as the user is concerned they all function in this same way. They do

http://dx.doi.org/10.3362/9781788530705.006

not need direct sunlight; in fact, some modern PV technologies function happily using artificial light. They operate autonomously, with no moving parts. Moreover, there are no visual signs of operation, no emissions of any kind, and no sound is produced – not even a hum. The seemingly unremarkable PV module (visually not much more than a metal-framed piece of glass) will provide electricity to a connected battery whenever there is daylight.

The two types of PV system

So far, so impressive. But, as with batteries or engines, the modules are useless on their own. They need to be incorporated into a system to provide any value. PV systems are collections of discrete electrical components, which draw and condition the PV-generated energy into one of two distinct forms of usable electricity: Either *alternating current* (AC) for powering mains equipment, or *direct current* (DC) for charging batteries. These two generic types of PV system are commonly referred to as 'grid-connected' and 'off-grid' (the word 'grid' referring to the mains or utility electricity network).[1]

The economics of these two systems are very different. The value of grid-connected systems is principally driven by the local cost of utility electricity. For stand-alone systems (remember, they are basically battery chargers) it is the value of the application being powered and the cost of the alternative energy solution for that application (if there is one) that quantifies its purpose.

This book is primarily concerned with stand-alone PV systems because they represent the greatest value for energy-poor and disadvantaged communities. I also believe that they represent the greatest commercial opportunity within the developing world energy sector – particularly for initial consumers of unreliable grids.

We will describe grid-connected PV systems below in moderate detail, primarily to emphasise the important differences from being 'off-grid', but also to highlight how the value of PV technology is being undermined. The PV industry is heavily focused on large installations that feed into the grid. Given what we have explored earlier in Part One about the limitations of the grid itself, this is of course a major issue for the future of energy in the developing world and industrial world alike.

Stand-alone PV represents the greatest commercial opportunity within the developing world energy sector

---

Grid-connected PV systems

Also called grid-tied systems, or just 'grid PV', these systems feed directly into the mains electricity network. They produce electricity only during the day, and there is no energy storage.[2] They are designed for maximum *annual* energy generation, which means the maximum revenue from generated electricity.

PV modules are installed on the roofs of homes, offices, commercial buildings and warehouses; and on numerous open spaces from farmland to deserts and even water. The PV rated capacity can range from a few hundred watts (Wp) to utility scale sizes of over 100 MWp (Megawatts, or millions of watts). The 'p' stands for 'peak', because PV modules are rated according to what they are capable of producing, not what they actually produce (more about this later). Meanwhile, installations typically receive subsidies: either for the capital cost of the installation, or for the green electricity they generate (from *feed-in tariffs*); in some cases, for both.

System operation

Grid-connected systems operate as follows. When daylight reaches the surface of the PV modules they produce DC electricity, which is fed to an *inverter*. The inverter converts this DC electricity into AC, conditions it to a specific voltage and frequency, then feeds it into the mains electricity system of the building or area. There is no change to the operation of any appliances within a building where PV electricity is generated. (In many installations, such as offices, the occupants may be unaware of a PV system on the roof.) If mains electricity is not present, or if it is the wrong voltage or frequency for whatever reason, the inverter disconnects the PV from the grid. Inverters are therefore essential components, and there are national and international standards governing their functionality and safety.

The value of grid-connected PV electricity

During daylight hours, the energy from a grid-connected PV system is either used close-by (for example, in the same building) or it is exported to the grid. When it generates more power than is being consumed, it exports any excess to the grid. In cases where the PV is not generating enough energy to meet demand, such as on cloudy days or at night when it is not generating anything at all, the consumer draws electricity from the grid as usual. To reiterate, these processes are fully automatic and no user intervention is required.

Revenue is earned for electricity which is exported or fed into the grid. The amount earned for each unit of electricity is commonly known as the *feed-in tariff*. Sometimes, revenue is earned for all the electricity actually generated, regardless of where it is consumed.

Out of necessity, PV modules used for grid-connect applications come with operational warranties of 20 years or more. The initial cost of the system is usually paid back within 10 years through feed-in-tariff earnings. Thereafter, the system is earning strong revenue. This scale of profit is what has driven the development of the grid-connected PV sector. For obvious reasons, subsidies and/or feed-in tariffs need to be guaranteed for the life of the payback model (which is not popular with governments).

Systems are designed to produce the maximum energy possible over the year. In somewhere like Northern Europe, over 70% of annual generation potential might come from the five summer months, so the PV modules are installed at a specific angle and orientation to take advantage of this.

I will reiterate the point that grid-connected systems are not backup power supplies; nor are they necessarily heroes of cleanliness or sustainability. If the mains grid is not present, or is not operating correctly (within voltage and frequency limits), the PV system will shut down by law. Nearly all utility grids around the world are dependent on coal, gas, oil or nuclear fuel for base-load generation. All grid-tied PV (and nearly all renewable energy) systems are therefore dependent on these fuels.

Certain improvements – battery storage, distributed generation and 'smart grids' (real-time monitoring and control) – are steadily increasing the percent-age of renewables that grids can accommodate, but the critical base-load still needs to be present.

Categories of grid-connected systems

Grid-connected systems can vary in the way they are installed. Alternatively, they can be categorised by their size or by the way they are funded; these two being closely linked. In terms of size, both electrical and physical, there are five broad categories.

- **Building integrated PV,** commonly referred to as BIPV, is where the PV modules replace part of the building fabric – typically roofing tiles, windows or shades (louvres/brise soleil).
- **Residential or domestic installations** are the most common sys-tems and range from a few hundred watts to a few thousand (kWp). They are also considered to be the future for grid-connected PV because they are a versatile form of distributed generation and as such can reduce demand from (mains grid) central generation and transmission.
- **Small commercial** systems range from a few kWp to 10s of kWp. These are also strong forms of distributed generation.
- **Large commercial** systems are approximately 100 kWp to over 1 MWp. Although valuable as distributed generation, their capacity requires care-ful consideration concerning their impact on the local grid system.
- **Utility scale** systems can be over 100 MWp. As recently as 2005 there were only a few instances of this scale of installation. Today there are hundreds of systems in the tens of megawatt scale. These are the pre-ferred installation type for PV manufacturers because of the volumes of modules used, and it is no surprise that the PV industry has pushed for larger and larger installations. However, they require significant plan-ning, installation, operation and, most critically, financing; all of which pose challenges.

'Off-grid' PV systems

The phrase 'off-grid' doesn't just refer to a distance away from grid electricity: it means a system which has no involvement whatsoever with the grid. It was the PV industry itself that coined the common term for these battery-charging systems as 'off-grid'. By referencing the grid in this way, I feel that we undermine the perceived capabilities of these generic energy solutions, and limit understanding of them.

Originally, all PV systems were off-grid. Satellites were the first examples, there being no realistic alternative for powering them. Calculators became the most well-known because they were the first to bring PV into the hands of consumers and the general population in the 1980s.

When designed correctly, off-grid PV systems can be the most reliable form of electricity available from a single technology

---

System operation

Nearly all autonomous PV systems have three main elements: PV, battery and charge controller. The charge controller is primarily to protect the battery and maximise its life. Depending on the requirements of the application, the battery can be sized to supply power not only though the night, but also during periods of bad weather and through the seasons. (There are a few applications that do not employ a battery, such as water pumping and ventilation, but the vast majority of systems are based around energy storage.) Electricity can be drawn directly from the battery as DC (the form used in vehicles), or through an inverter to produce AC – mains-equivalent electricity.

The value of stand-alone PV electricity

Stand-alone systems do not export any energy. Any electricity that can be generated by the PV is fed into the system battery, is used directly by equipment in the system, or it is not used. Here, the PV electricity itself is not of direct value; rather, it is the *economic benefit* or *the service this electricity enables* that is valuable to the user. The value of these systems lies not in their own cost. Instead, it is the comparative cost of the *alternative* energy solution that defines the value of a stand-alone PV system.

Example 1: Telecommunications
For example, if a telecommunications transmitter needs to be located on top of a mountain, it could cost hundreds of thousands of dollars to install a mains grid supply. It is unlikely that a diesel generator could be refuelled regularly; upon which, PV becomes the most valuable solution – even if the supplier is making several hundred per cent margins. This is exactly how the terrestrial

PV industry came to be. (The telecoms transmitters are a real-life example, installed in Oman in the 1970s.)

Large, fixed location systems can be installed almost anywhere, and their tangible value is significantly enhanced by their reliability. When designed correctly, stand-alone PV systems can be the most reliable form of electricity available from a single technology. Stand-alone systems designed for long life and high reliability do not usually require subsidies because they have the lowest lifecycle cost.

Example 2: Solar calculator

In products such as calculators or watches, little solar PV cells are incorporated as an improvement to the overall value proposition of the product, rather than an essential enabler of the service being provided. The tiny panel generates very small amounts of charge for the device's internal battery. This allows a small, low-cost rechargeable battery to power the product for many years.

The PV and small rechargeable battery are used because they are cheaper and offer superior value than a larger non-rechargeable battery (which would yield one or two years' operation, at best).

Example 3: Street furniture

Over the last 15 years or so, 'street furniture' has been a growing market for off-grid PV. Another strong value proposition, it refers to installations along roadways and public areas, and includes bus shelters, amenity lighting, emergency roadside phones, parking meters, advertising, waste bins and more.

To illustrate the value of PV, take the example of a parking meter. A PV-powered version of this meter costs significantly more than the regular type; yet there are thousands of 'solar-powered' versions around. Why is this? Standard meters require very little electricity to function, but a mains connection has to be made nonetheless. (Alternatively, powering the meter from a battery requires frequent replacement; a significant logistical cost where hundreds of meters are deployed over a large area.)

To install a mains-powered meter, you have to dig up the street, extend a mains utility cable to the site, pay the local utility a connection fee, enter into a payment contract for the electricity, and inspect the installation periodically to comply with safety legislation. But a solar-powered meter simply needs bolting to the ground; and it works, straight away, costing very little to run. The value of solar for this type of application lies in the *avoided* costs – which are both financial and in terms of time. If the meter does not have power, it is not earning you any money. This value proposition applies to all street furniture requiring power.

Solving problems better than alternatives

It is critical to understand the central concepts of these three examples – the telecoms tower, the calculator and the parking meter. The amount of electricity generated by the PV is not important to the customer, nor is the efficiency of the PV; nor even the specific technology or cost. The requirement for the customer is that the PV system as a whole solves a problem *more cheaply and easily than any alternative*. This is the key to understanding where financial profit can be made. Stand-alone PV systems are not priced in terms of their cost, plus a margin. They are valued by the cost of the alternative minus an attractive saving to the customer. The challenge is to represent and convey these savings adequately and comprehensibly.

Categories of systems

Within the abundance of stand-alone PV applications, there are thousands of solutions in a variety of different sizes. This makes categorisation a challenge, and it is one of the reasons for the common confusion regarding what these systems actually do.

I use the following three broad categories: *portable and consumer products*; *regional solutions*; and *bespoke systems*. Even within these groups you will find diverse applications and value propositions. But they provide clarity when considering the way their respective PV solutions are designed, supplied, supported and marketed. These are the most important aspects of their widespread deployment, particularly in energy-poor regions. Meanwhile, these categories also generally represent the solutions by size, both electrical and physical.

Portable and consumer products

Portable and consumer products typically have the PV and battery embedded into the product, and the load equipment (the bit actually consuming the energy) is usually either incorporated or attached by wires. Such products include: watches, calculators, garden lights of all kinds, radios, toys, water fountains, consumer battery chargers, bags (with various functions), keyboards, headphones, mobile phone chargers, torches, walkway markers, LED lanterns and fans. These products number in the hundreds, if not thousands.

Very few of them work effectively. Performance is entirely dependent on the user for both energy generation (how much sunlight the PV is exposed to) and consumption (how the device is used). This is also true for devices like LED torches, where the load is known. Even with good instruction or even training for the user, poor performance and low reliability are common. These products cannot have predictable performance in real world conditions; they can only be supplied with guidelines of their potential performance.

Portable and consumer products sell in high volumes worldwide. Very few of them work effectively

---

The cost of the product depends on the capacity and type of the PV and battery. Lower quality or capacity PV will mean a cheaper product, but of course less energy generation, and lower performance. Smaller batteries are cheaper and lighter, but hold less energy. Although they are often only designed to work effectively in sunny locations, these products sell in high volumes worldwide, under market pressure to be as cheap as possible. They are regularly promoted on the strength of their best ideal performance, rather than on reasonable expectation.

Regional solutions

Regional solutions can be thought of as 'one design for multiple locations'. These designs are produced in modest volumes (when compared with portable and consumer products) and are often deployed as a 'solution in a box'. PV modules can either be discrete, or incorporated into the product; the battery and the load equipment likewise. Regional solutions include street furniture (bus shelters, bus stop lighting, parking meters, emergency phones), electric fencing, refrigeration, television, water pumping, Wi-Fi hubs, domestic solar home systems,[3] street lights, navigation lights, all sorts of 'rural electrification',[4] professional outdoor solutions, emergency shelters, tents and a whole range of monitoring devices.

The user device or appliance is typically supplied and so the electrical load is a known quantity from the outset. While the degree of use can vary enormously, in general the performance is defined by the location in which they are installed (climate conditions) and the solar quality of the installation (the amount of shading on the panels, their orientation, dirt and so on). To accommodate the different climates around the world, one central design may be supplied in a range of sizes – with higher PV and battery capacities for cloudier climates. Because these systems are designed for assumed levels of sunlight, they should only be sold into defined geographical regions (hence the category name). Unfortunately, this is often not the case. 'Europe', for example, is a single commercial market, but there is a big difference in sunlight between Scotland and Greece.

The value of these solutions is usually rooted in the offset cost of an alternative, as we saw for our parking meter example. They are often designed for 24 hour-per-day energy supply; and when designed and installed correctly, they can perform reliably for many years without maintenance. A high proportion of potential developing world solutions fall into this category. For the 2 billion people without access to reliable energy services, regional solutions hold the greatest commercial opportunity.

Bespoke systems

Bespoke systems can be thought of as 'one design for a single location'. They are usually designed for a known climate, a specific installation site, a defined electrical load requirement and an agreed level of reliability. Their varied applications include satellites, telecommunication towers, offshore oil and gas platforms, pipeline protection, water management and irrigation, remote research stations, shipping container systems, boats, safety systems and site security. They also encompass a range of remote building and community electricity supplies.

Bespoke systems are the largest and most difficult stand-alone PV solutions to design, but the most valuable, because of their reliability, long operational life and low maintenance requirement. They are typically designed to provide energy 24 hours per day, every day, for many years. For industrial contracts, an operational life of at least 15 years is commonplace, with batteries lasting from 5 to 10 years.

By their nature, such systems require significant design work, and are deployed in relatively low volumes. But in the absence of viable alternatives (as seen with our Oman telecoms example earlier) they are immensely valuable. As I will mention repeatedly in this text, when designed correctly, bespoke autonomous PV systems are the single most reliable form of electricity available.

Bespoke stand-alone PV systems come into their own in energy-poor regions. In the absence of a utility grid (or a sufficiently reliable one) they can enable much of the essential infrastructure needed for social-economic opportunity – such as communications, education, healthcare and media.

It is encouraging that bespoke systems are becoming increasingly popular for supplying multiple domestic as well as commercial end users in the developing world. Unfortunately, 'mini-grids' (as well as micro- or even pico-grids) are typically promoted as alternatives to the utility grid, often supplying AC electricity. As we saw in Part One, retaining a reference to the grid, let alone trying to actually mimic the grid, is a barrier to efficiency and wider progress.

Different worlds

In all of the above categories, there are unsettling differences in the products and solutions for the industrialised world compared with the developing world. Large bespoke systems for industrial applications typically carry 10-year or longer operational warranties, whereas the equivalent system in the developing world (sometimes using the same components) can last as little as a year. For the regional solutions, the highest volume application is a 'solar home system' for a rural energy-poor family; where in the industrialised world it is likely to be a monitoring device, or an outdoor pursuit application (say, for camping).

But it is the portable and consumer products which expose the greatest gulf between rich and poor. In the developing world, products are governed by cost, the designs fairly basic and the functionality borderline acceptable for the needs, at best. For the industrialised world, the emphasis is on design, with the PV far too often being a gimmick – in many cases the products have removable batteries, mains charging facilities, or both. This is because 'off-grid' consumer products are not really needed by most people: we all have access to the mains nearly everywhere, while our vehicles serve as charging devices, should we stray from the grid.

For the 2 billion people without access to reliable energy services, regional solutions hold the greatest commercial opportunity

A summary of PV systems

Grid-connected systems

- The five general categories of grid-connected PV systems are building-integrated PV, residential, small commercial, large commercial and utility. All function in the same way, feeding PV electricity into the utility grid. They can only do this during daylight hours; and only when the grid is present and electrically stable.
- All grid-connected systems share the same value proposition: they produce clean electricity, and money is received from someone (government or utility company) for that energy. PV is installed to maximise annual generation for the given climate. This requires orientation and inclination of the modules to maximise revenue.
- The capital cost of the installation, which is proportional to the capacity of the PV, is paid back over up to 10 years, then money is earned for the rest of the life of the system – usually over 20 years in total. Subsidies and feed-in tariffs need to be guaranteed for the life of the payback model.
- Grid-connected installations are a good investment if you have the capital and accept the long payback duration, and if you have confidence that your utility grid will be active and within operational tolerances for the vast majority of the PV system life.
- The larger the installed system, the greater the financial benefit over the life of the contract. The PV industry has therefore pushed for greater scale, despite the challenges.

I believe, as do many in the PV industry, that relatively small, residential and commercial installations will become the dominant form of grid-connect system. These offer a variety of advantages: creating jobs, cutting emissions, stabilising national grids and reducing foreign oil dependency. But as we have discussed, these motivators are slow to take effect.

Stand-alone systems

- Stand-alone systems differ, not only from grid-connected versions, but often from each other. There are common factors – such as PV charging a battery – but they can also be very distinctive.
- Value is not associated with the PV's capacity (as with grid-connect systems).
- *Portable and consumer products* are highly dependent on location and method of use if they are to even come close to the stated performance claims of their suppliers. *Bespoke systems*, on the other hand, have a historical pedigree for reliability, long life and low maintenance.
- The costs of *bespoke systems* and *regional solutions* are closely associated with the local climate and the load energy or service required. For *portable and consumer products*, this relationship is far less evident.

- The value of a stand-alone system cannot accurately be expressed in relation to its own expense; but rather in terms of the saving made or the benefit gained, compared with an alternative method. This is a difficult concept to communicate, which explains why relatively few people and organisations understand what they can do, and how valuable they are for future energy.

The scope of stand-alone PV solutions

There are hundreds of uses for PV systems and thousands of application sizes and formats. After all, they are electricity generators without the need for fuel; they can be almost any size and can work pretty much anywhere on the planet. This versatility is critical to PV's role in addressing global energy requirements.

Even veterans of the sector have difficulty conveying the scope of existing stand-alone PV applications, let alone those yet to be implemented. It is simply not possible to categorise each application and the value it represents clearly, because – even for a given product, such as a solar light – that value is so specific to the user, the location and the application. An altogether more valuable exercise is to convey the true scale of *potential*. To this end, below is my attempt to present an appreciation of scope to spur the imagination of what a versatile and autonomous energy solution can enable.

Generic services and sectors of society

A good starting point is to consider the generic services for which stand-alone PV is most commonly used. From this broad base, we can consider the more recognisable applications for these services.

I don't intend to delve too deeply into the specifics of either the services or sectors in which they are used. I have also excluded the dozens of more specialised applications of stand-alone sustainable energy. However, it would be amiss to leave these out altogether, so a list is included to offer an overview of more specialised applications listed by sector.

The most common generic services enabled by PV

These are the primary energy services that PV is ultimately used to enable. This does not list the devices or equipment; rather it illustrates the base level with which to categorise the applications, thereby to assess the relevant technologies and devices.

- Lighting
- Phone charging
- ICT (information and communications technology)
- Monitoring

- Media and entertainment
- Refrigeration
- Device charging
- Battery charging
- ICT peripherals

Some of these services in themselves are vast in their scope, and you can categorise further to several levels of abstraction. For example, within the generic service 'lighting', we can list: portable navigation (torches), kerosene lamp replacement, personal reading/craftwork/food preparation, domestic rooms of all sizes, public and commercial buildings of all sizes, public services (toilets, emergency phones), medical lights, street and amenity lighting, security and safety. Further still, 'amenity' lighting as a single category includes: walkways, car parks, gathering and trading areas, healthcare waiting areas, building security, perimeter areas and hazardous areas (holes, cliff edges, water, wildlife, etc.).

Sectors of society where these services can be utilised

To establish the scale of commercial possibilities, consider these 'standard' energy services against the many different sectors of society in which they can be utilised. These include the following:

- Healthcare
- Education
- Communications
- Business and commerce
- Personal devices
- Water
- Community energy
- Infrastructure
- Domestic services
- Shelter and enclosures
- Agriculture
- Transport
- Marine
- Security
- Recreation
- Outdoor pursuits
- Consumer and lifestyle
- Gadgets, gifts and toys
- Military
- Indoor

Example stand-alone PV applications by sector

The list below offers an appreciation of the more specialised applications. Note that several applications appear in multiple categories. These may have similar central functionality but they usually have different features such as environmental resilience, security (anti-theft), ruggedness (for logistics and installation) and so on.

Oil and gas infrastructure	Monitoring, control, safety, corrosion protection, private communications
Telecommunications infrastructure	Satellites, large remote repeater stations, local transmitter receivers
Communication devices	Mobile and satellite phone charging, internet access, international phones

Monitoring	**Air**: greenhouse gases, pollution, temperature, humidity, wind speed, **Environment**: noise levels, traffic volumes, **Water**: rainfall, contamination, flow rates, levels
Homes (developing world)	Lighting, IT, entertainment, communication, refrigeration, water purification, security, device charging
Healthcare	Portable or fixed refrigeration (vaccines, blood, organs), lighting, monitors, water purification, incubators, centrifuges, sterilisation, IT, cameras
Rural electrification	Lighting (domestic, community, medical and educational centres), water (pumping, filtration and irrigation), electric fences, security
Transport infrastructure	Road, rail and waterway lighting, signalling, safety signage and control, monitoring, temporary/portable traffic control
Marine	Buoy lights, lifebuoy lights, sensors (e.g. tidal, wave and tsunami), location devices, communications, radio, device charging
Security	Barriers, perimeter fencing, surveillance, motion sensors, door and window sensors, lighting
Media and IT	School and college computers, TV, radio, educational films, camera charging, video equipment charging or powering, etc.
Leisure and outdoor pursuits	Caravans, boating, holiday homes, camping: refrigeration, security, lighting, insect repellent, device charging, etc.
Personal and consumer	Calculators, phone and PDA charging, torches, watches, personal security, radio, consumer battery chargers
Temporary shelters	Tents, marquees, portable cabins and shipping containers used as medical facility, resuscitation tents, rain shelters, sun shelters, insect barriers (particularly mosquito), lighting, cooling, location devices, phone chargers, IT and medical equipment, etc.
Indoor	Smoke alarms, air fresheners, active price tags, security tags, gas sensors, door locks, computer peripherals (keyboards and mice), etc.
Outdoor (industrialised world)	Amenity lighting, security, monitoring, advertising, electric fencing, emergency phones, pest control, location devices, street furniture, etc.
Military	Soldier, remote command posts, perimeter security, communications, location tags, GPS, IT equipment, night vision, lighting, etc.

Off-grid PV and personalisation

I have specialised in stand-alone PV for over 20 years. In that time, I have seen the promise of what it can achieve continuously reworded and rebranded. Back in 1999, PV technology was promoted from the environmental angle for the clean electricity it generates. As the years rolled by, the rhetoric shifted to fuel offsetting and the subsequent benefits of reduced overseas dependencies. There followed a couple of panic periods. Here, global oil prices spiked and the commercial benefits of PV systems were discussed, largely inaccurately. Since then, more considered value propositions have been forthcoming; for instance, how a PV array could increase the value of your home; or the positive 'green' statement PV can make for a commercial organisation.

Today, PV systems are widely acknowledged as demonstrably cleaner for the global environment than any fuel-based energy generator. And stand-alone PV is now accepted as a healthier, safer and more economically sound option for those without access to modern energy services. But neither the benefits nor the rhetoric have ever been enough; and these systems are still marginalised. This is what eventually led me to the 'Personalised Energy' perspective: the need for the user to be involved with the system and value the service it is providing; and especially the need to spread knowledge and awareness effectively.

Hopefully it is becoming clear that stand-alone PV solutions hold enormous value for energy-poor regions – meaning people without reliable energy services. In their many forms, these solutions can meet the needs of the 'hardware' element of 'Personalised Energy'. They can be engaging, intuitive to understand and simple to use. As such, they can reduce the burden of what the 'soft' element – knowledge sharing – needs to convey.

However, stand-alone PV comes with specific demands on the user and it also comes with historical baggage. Next, to address these and other challenges, we will investigate how to convey information in a universally accessible, culturally neutral manner, and the value that this brings.

Endnotes

1. The terms 'PV' and 'solar' popularly tend to be used interchangeably, describing both PV technology and PV systems. Also, due to the prominence of grid-connected PV applications, even those within the industry commonly use the terms 'solar' or 'PV' to mean 'grid-connected system'. For the avoidance of doubt, I am referring to autonomous PV systems (which I prefer to call 'stand-alone' systems), unless I state otherwise. (I will explicitly state 'grid-connected' or 'technology' when discussing those issues.)
2. Domestic storage products are available, but at best they improve economics and offer a few hours' backup. Grid-connected PV with storage constitutes only a very small percentage of installations (though this is likely to grow significantly in domestic applications).
3. A generic term for self-contained, domestic PV solutions (see Chapter 9 on quality and value).
4. A developing world term which covers lighting (domestic as well as public), water (pumping, filtration and irrigation) and security (see the table of example stand-alone PV applications by sector in this chapter).

CHAPTER 7
Universally accessible visual media

Abstract

Making energy a personal affair means confronting scale and diversity – massive numbers of people living in diverse circumstances.

In this chapter we describe how visual media can fulfil the knowledge sharing requirements of our strategy.

We define our Visual Instruction System (VIS) through discussion of the many inherent restrictions on our media, the core requirements of neutrality, being inviting and engaging, and the messages we need to convey for PV systems. Visual information will dramatically increase the value that users gain from their products, and bolster the reputation of the generic technology.

If we can help people understand autonomous PV and its potential, people will not only learn the range of existing applications; they will learn how to adapt them for the greatest personal value and think of new ones with distinct value for themselves and those around them.

Utility grids conceal their methods of operation from us. Given their significance to society, it is strange to think that they require almost no understanding of their function; thus, they hide the importance of user interaction with energy. However you choose to abuse the grid or waste its output, it keeps on giving; well, for industrialised societies at least. In the developing world, such inefficiency imposes severe technical and economic limits to how many people can be served.

My early solar life was mainly preoccupied with large industrial, bespoke PV systems. These were specifically designed to operate without human input, their very value based on it being prohibitive for anyone to visit the systems. Only when I became involved with urban projects did I realise the critical importance of users. For stand-alone PV solutions, any given operator has to understand how the system functions, to ensure effective and efficient performance. I realised that this was true of most battery-enabled equipment. You can't get the most out of your smartphone battery unless you understand which functions are most energy-intensive. The same is true of portable computers and other battery-powered devices.

When it comes to stand-alone PV, the critical demand for communication is there, in plain sight. If the user does not understand how PV operates – the PV panel has to be in sunlight for effective generation – it will fail to perform to its potential, no matter how good the product or solution might be.

http://dx.doi.org/10.3362/9781788530705.007

Communicating with people is critical to the success of stand-alone PV, at a local and personal level. If you recall, the principles of 'Personalised Energy' emerged because stand-alone PV solutions require the user to engage with the system in order to reap its rewards. Here they are again:

- We need to involve the user in the generation and consumption of their energy.
- We need the user to value the electricity they have at their disposal.
- We need to convey knowledge and understanding of all of these areas.

When we consider the dramatic consequences of fossil fuels and fuel dependency, these are valid principles which apply to all energy service solutions. We will see that it takes more than mere cables to make this connection.

Why do we need visual media?

Making energy a personal affair means confronting scale and diversity – massive numbers of people living in diverse circumstances. Our broad target – those with unreliable grid electricity or limited access – number in the region of 1 billion. The official figure for people without grid electricity is 1.3 billion and we must include this most essential group in any strategy for energy provision.

Within this giant demographic, any information we use to address energy services needs to be universally accessible. This constrains our options considerably in the media we use. For example, around 800 million people in the world are illiterate. This would appear to be a show-stopper: we cannot communicate in writing with people who cannot read a single language or even a set of languages between them. Even if we were to find an efficient way to translate text, the written word is simply not engaging or succinct enough for our needs. (And how many literate people read the instructions for their electronic devices these days?)

Furthermore, with over 7,000 living languages in the world, communicating using audible speech is not an option. Even if you discount the localised tongues, assuming a primary language for a given region, the number is still prohibitive. You can generalise further still (ignoring some broad cultural assumptions along the way) within a baseline of, say, 200 languages. But this is still too many to be practical, and you will already have excluded the fringe populations, the rural and the needy – the very people we aim to reach. If we cannot use written text or speech, we will have to communicate visually.

Some challenges of visual media

Turning our attention to the visual realm, we find again that scale and diversity place conditions on how we communicate. Without embarking on a scholarly account of Communication Theory (a field too vast to summarise in these pages), we can acknowledge a few generic principles which act as a foundation: *simplicity* (the need to say as much as possible with the least number of visual cues; or the need to organise elements into a comprehensible whole) and *consistency* (establishing a single, universal methodology and style of communicating). Our current task, meanwhile, is to communicate rudimentary technical knowledge across many civilisations; and it poses more specific challenges. Our target audience lives in well over one hundred countries, demonstrating varying degrees of literacy and technological proficiency, and representing a diversity of culture, religion, age and gender.

Neutrality

In May of 2013, I hired a driver to travel from Nairobi, Kenya across to Uganda. Apart from the most dangerous driving I have ever been exposed to in my life, the driver constantly made reference to how everything bad could be attributed to the corrupt Kenyans, while everything good was from Uganda. (Being arrested by Kenyan police for speeding didn't help his attitude, or our progress.) A few days later as we neared the Rwandan border he turned his attention to them: the pineapples from Rwanda were awful and nowhere near as juicy and tasty as those splendid examples from his motherland.

> Our target audience lives in well over one hundred countries, demonstrating varying degrees of literacy and technological proficiency, and representing a diversity of culture, religion, age and gender

People have prejudices and biases. That we cannot see through the false barriers we have created among ourselves is a weakness of our species, but they are there nonetheless. The biggest separators of people are gender, skin colour, religion and nationality. This poses a real problem when considering how to reach people with information.

We must consider neutrality in our communications, and its scope is broad and subtle. Depicting women as knowledge givers, for example, while thankfully no longer controversial in many societies, may nevertheless fail to engage with patriarchal cultures. An affluent white man giving advice to poor black people may appear inappropriate. The best material we can produce will ultimately fail to connect in Muslim countries if they contain assumptions or symbolism of Judeo-Christian cultures. Put simply, we need to be neutral: no cultural, religious, ethnic, gender or age references in our knowledge sharing material. In order to avoid being off-message, the challenge is to make 'no message' beyond a primarily factual or instructive one.

The requirement for neutrality largely precludes the use of real people in our media. We cannot show skin colour and for similar reasons we cannot show real world backgrounds. However, we will need to show how the user interfaces with our hardware and how to carry out such actions as operation, cleaning and maintenance; so this will mean showing hands or depictions of people.

Symbolism and colour

Due to the diversity of people, the general design concept or schema must be as neutral as possible while maintaining relevance. Certain colours, animals, objects and symbols are prohibitive to some cultures. The colour red, for example, can symbolise love, passion, danger, communism, witchcraft or death depending on where you are in the world. Cows, dogs and cats have a sacred status bestowed on them by various religions. Meanwhile, in addition to the widely revered (and also divisive) symbols of association such as the crucifix or the six-pointed star, there are other more benign symbols which are very much the result of industrialised societies, whose widespread use belies their limitations. The western pink/blue gender distinction, for example, is mere convention, and a recent one at that. The use of ticks or crosses, green-amber-red notation or thumbs-up/thumbs-down signs are modern-world gestures which can denote basic positive/negative, stop/go meanings; but we cannot assume that they travel well. (As an aside, visual media research also highlights the *accessibility* problem when using colour symbology, noting the large proportion of colour blindness occurring in the green and red colour band, among certain populations.)

Due to the diversity of people, the general design concept or schema must be as neutral as possible while maintaining relevance

---

Global or regional material

Under such restrictions, it would be tempting at this point to just admit defeat, and accept that universal, culturally neutral material is simply not possible. Many of the artists, animators, photographers, teachers and media specialists I have spoken to have stated this; maintaining that our universal requirement must be divided into regional formats.

I have researched this approach. What I have found is that subdivisions such as 'Asia',' India', 'Africa', 'South America' and 'others' are largely (if unintentionally) based on skin colour; and in the many examples I have reviewed, there is a definite assumption on the basic learning ability of each group. I fundamentally disagree with segregation of this type. The people in these different areas do not have a limited ability to learn; they have limited resources and disproportionately time-consuming basic requirements of everyday living.

But one thing amid this research particularly stands out. The 'other' category in the list above basically stands for white people – USA, Europe, Australia and so on – and despite the obvious assumptions of learning ability, much of the material for knowledge dissemination is focused on trying to make it eye-catching with infographics and bold slogans, or straplines.

We have noted how our material needs to be inviting, engaging and non-threatening, while avoiding any cultural, religious, gender or age references. Moreover, in industrialised societies, media production functions on the assumption that its consumers are busy and have limited attention spans.

This requirement is identical for the consumer society and the energy-poor world: for the busy urban executive (who, let us say, makes decisions about energy service provision) and for the mother in a semi-urban developing nation who needs convenient access to modern energy services (and who, as it goes, would love to have the spare time to engage with what we are presenting).

Examples of visual media

Reassuringly there are precedents in this field. IKEA is the best-known and most widespread: all their assembly instructions are in black and white, simple line drawings with no words or cultural references. (I hear the screams of the many who have found these instructions a challenge, but they do work...) Luckily, we do not need to illustrate detailed furniture assembly; we only need to convey a handful of fairly basic principles.

A growing number of global brands are no longer printing instruction manuals in 20 or more languages. This is largely due to the rise of newly affluent markets in Asia, and their variances in technology awareness, literacy and dialects. Logitech and Bose, for example, have image-only user instructions now, while others such as Dyson (vacuum cleaners) and Petzl (head torches) use images supported with text. Many other high-volume manufacturers provide 'quick-start guides' exclusively in picture format and support this with online-only written manuals. Whether images are used exclusively or otherwise, the value is obviously apparent to these large organisations.

However, few companies are actively pursuing this approach for the developing world, where it is most needed. Grundfos (which produces high-efficiency water pumps, including some supplied with PV) is a notable exception.

The Sundial example

Let me describe a real example of why this is so important. Between 2010 and 2013, I worked with a Chinese manufacturer to develop a range of stand-alone PV products targeted at energy-poor regions (a courageous strategy at the time). Our first product was a solar light, named Sundial: an award-winner which was especially popular with the middle classes in urban areas with unreliable mains electricity. The product was distinguished by value and quality rather than a competitive price point; the entry-level Sundial was in the region of $35 versus typical competitor products selling for around $20.

At first, nobody bought the Sundial (and I don't mean figuratively). Then, when the sales team in Kenya started highlighting the product's features (2-year warranty, 5-year life, rugged construction and remote control), customer preferences changed in our favour.

Next, we applied some basic principles of visual information to the product packaging and the sales sheets. Crucially, the sales feedback confirmed increased sales without incurring further sales team costs. There was a marked improvement in sales volumes and margins.

Economic reasons ultimately prevented us from realising the full potential of visual media to support the Sundial. However, we had demonstrated
beyond doubt that the visual approach was an effective sales tool. Revealingly,
image-based sales brochures started to be produced by our competitors.

No company yet has fully grasped the concept of visual media with universal accessibility. It is a difficult concept to execute well, so there is still a
dependency on words and sales staff. But the appearance of packaging and
user instructions is changing. People are drawing greater meaning from product boxes, the products themselves and from some of the sales literature; and
this trend bodes very well for the reputation of PV products, as well as the
generic technology.

Abstract concepts: describing time and change

So far, we have assumed an emphasis upon static media, for example on packaging, or instructions. But fundamentally, our goal is to promote a tangible
sense of gain or achievement, in order to bond a user with the benefits of
the device. The first challenge is to find universally recognisable symbols.
For 'solar' products the basic concepts are thankfully easy enough to identify: the sun, the moon, clouds, stars and facial expressions. There are a few
genuinely ubiquitous objects we can use: birds, dust and dirt, water and food.
Things start to become a little difficult where we attempt to show subjective or
abstract concepts, such as time, value, versatility, safety and security.

Of particular importance is the concept of time. Firstly, remember that we
are aiming for more than a protocol for removing a product from a box and
pressing an 'on' switch. Also, in the world of PV we need to convey change,
over hours, days, weeks, months and years. This might seem simple at first,
but consider that we cannot use clock faces or watches alone. We cannot use
numbers. We cannot depict the Earth orbiting the Sun, as this is not universally understood. The length of a day varies throughout the year and/or from
one place to another. Seasons are different at the equator and in the tropics.
Crops grow at different rates in different places. The list of prohibitive imagery
goes on.

As an exercise, think of how to show that one product lasts 5 years while
the slightly cheaper competitor product lasts less than a year. What is it that
absolutely everyone can relate to that takes months and years? (One answer is
at the end of this chapter, if you're impatient.)

Animations and storyboards

While print media has undoubtedly the furthest reach (being self-contained
and requiring no electricity), it has its limitations. Trying to show, for example, that although a solar LED light costs 2 months' wages, it will last 5 years
without any operating costs, is not easy. This is before comparing it against a

kerosene lamp that costs one week's wages every two months; not to mention the LED light's health and safety advantages against the flickering, smoke producing naked flame of the lantern.

Animations are capable of conveying the more abstract or complicated aspects of our energy services, including value, quality and time. While standard video may suit a limited number of purposes (such as showing time-lapsed user interface functionality), our requirements for neutrality and universal accessibility steer us in the direction of moving images which are pictographic, rather than photographic. Colour and shading schemes can indicate positive or negative scenarios, or mood changes (the brightness and sharpness of images indicating good and bad, happy and sad).

Animations are proven platforms for explaining concepts as broad as the structure of the atom to the scale of the universe. We may be barred from using familiar narration or text subtitles to assist in the learning process; however, there are techniques that we can use to strengthen our central content.

People can by nature be nervous of the unknown, and our audience may not have seen animations or storyboards before. So any material we produce needs to be inviting and engaging while avoiding a threatening or otherwise alienating experience.

Foundation knowledge and value

Having discussed the pros and cons of our chosen media, what is the actual message? Stand-alone PV has a baseline requirement of the user: they must at least appreciate that the PV panel needs to be exposed to daylight; and unless the PV is integrated into the product, the PV must be connected to the battery unit. Two examples follow.

Simple mobile phone charging

Consider the most basic but essential instructions for a simple solar charger for a mobile phone. This is now one of the most common PV applications in the developing world:

- Lay the solar module in sunlight;
- Do not allow shading of the module during charging;
- Keep the phone out of direct sunlight during charging (under the PV module).

The sun and clouds are both universal objects, which can be represented in simple drawings. Positive and negative outcomes can be shown by a simple circular face, with a happy or sad expression. The PV panel and the phone are not universal but can still be idealised in a diagram in a meaningful way. Just three basic black and white drawings are sufficient to convey these essential messages.

Stand-alone PV

The basic knowledge elements we need to convey for generic stand-alone PV include the following:

- Locate the PV in daylight;
- Do not shade the PV;
- Connect the PV to the battery unit (if appropriate);
- Poor weather will reduce charging;
- For best results point the PV at the sun (the position of the sun changes throughout the day and year);
- For best results, the PV should be allowed to charge the battery all day;
- For best results, keep the PV panel clean.

We will want to extend these messages to show, among other things, a product's features, its performance compared with the charging conditions, how to take care of it, and what to do if it stops working. Most importantly, we will need to be proactive in reinforcing sales information, such as a product's financial value – especially as all of the cost is upfront. This will appear expensive to the uninformed customer, even though the product will cost nothing to operate over its lifespan (unlike almost all alternatives).

Am I stating the obvious? These examples may sound too basic, even unnecessary, but I can assure you otherwise. Throughout my engineering career I have been amazed by the incorrect assumptions people make about technology, and the false logic they apply. One of the most striking memories I have of this was in an office in Abu Dhabi many years ago, when I asked a receptionist what number I should dial from my mobile phone to reach her landline. I tried the number she gave me but it didn't work. She then suggested I stand slightly further away and try again. Meanwhile, I still hear people declare that it is better to leave a light on in a room for an hour than to switch it on and off for each momentary visit. The same is true of drivers who cling to the misconception that it is better to leave the engine running than waste large amounts of fuel starting the engine after a five-minute wait.

In very hot places, it is natural to want to keep everything shaded. Therefore, putting a PV panel in the sunniest (and therefore hottest) location can seem, on the face of it, counterintuitive. It is common for solar product suppliers to train rural villagers in the use and care of PV panels, but they often neglect to point out that there is no actual energy storage in the panel. It is by no means obvious that the panel needs to be connected to something (a specific battery) in order to be useful – especially when the conflicting golden rule for batteries is to keep them as cool as possible.

In very hot places it is natural to want to keep everything shaded.
Putting a PV panel in the hottest location can seem counterintuitive

---

I speak to many people within the developing world energy sector, NGOs and commercial organisations involved in the supply of PV products. All testify to a widespread lack of understanding of what solar PV is, or how to use it effectively. Even commercial projects reflect this. I have photographs of PV installations with metal aesthetic structures permanently shading the PV surface, causing up to 75% reduction in energy generation. I have seen at first hand many installations where PV is fixed facing away from the sun; behind a local feature such as a tree or roof element; or otherwise so dirty that you can no longer see the active surface. These installations involved a procurement process, a contract award, supposedly skilled installation and formal commissioning. They will have cost thousands, if not tens of thousands of dollars. (There is a PV system at the top of a renowned architectural tower in southern China, installed and promoted by a solar PV company. It is now referenced in a university course in renewable energy as a shining example of what *not* to do for a solar installation.)

However simplistic they may seem to a non-novice, the fundamental concepts of installation and a device's most basic functions *do* need describing. As we saw with our mobile phone charger, the value lies in conveying *enough* information to invite the user to interact correctly. The more knowledge we can convey the better, but we only need to focus on the basics to make a huge impact.

The reach of visual media

A Visual Instruction System (VIS) is complicated to devise and produce, but once the basic building blocks are established – how to represent time, culturally neutral characters, how to show good and bad – its very simplicity makes it suitable for a wide range of applications, and to make the most of contemporary digital media. Animations and still images can be propagated via email, websites and social media, as well as offline storage such as DVDs or USB memory devices. Visual information can be displayed on mobile phones, televisions (as commercials or from personal media), computers, tablets and projectors. They can be shown as the prelude to the main feature at the cinema. Printed media, while less versatile, is broader in its reach. Traditional product packaging, assembly and care instructions, environmental guidelines, and sales and marketing literature can be compiled in print form as leaflets, fact sheets comic books or storyboards. Combined, this allows us to reach a hugely diverse body of people: existing users of solar products, sellers and distributors, agents and partners, volume customers, governments, international donors, specifiers of energy services, and of course schools and educational centres.

VIS is powerful and possible. I spent six years developing the central concepts to address the basic needs for stand-alone PV; I have cracked many of the challenges to propagating essential knowledge for our autonomous energy services, and energy services in general. And, critically, I have seen this work in the field, supporting the Sundial solar light.

Visual information will dramatically increase the value that users gain from their products, and bolster the reputation of the generic technology. If we can help people understand autonomous PV and its potential, people will not only learn the range of existing applications; they will learn how to adapt them for the greatest personal value and think of new ones with distinct value for themselves and those around them.

We are all witnessing the global move towards visual and universally applicable media, as companies learn how to profit from the internet and reduce their costs. Just as the smartphone battery charge icon has helped wider understanding of battery management, these initial applications for the VIS on our stand-alone battery chargers will help develop the principles and understanding to more valuable and feature-based knowledge sharing.

(A baby is born. It lies in the glow of two solar lights. One year later it is crawling; beside it lies a non-functioning PV product. Two years later and the baby is walking. Year three, and the child is sitting in a chair with an appropriately proportioned adult for comparison. Year four, they are bigger with respect to the chair and the adult, and in year five they are eating and drinking on their own. The good product is operating in the background all the time and the poor one gathers dust and dirt. Happy and sad faces and variations in colour and shade denote the functionality of the equipment, and the basic sequence, over time, shows products lasting less than a year and 5 years respectively. I bet you could draw it.)

CHAPTER 8

The core value and poor reputation of 'off-grid' PV

Abstract

What do you gain from using a stand-alone energy service? And where does its value actually lie? Understanding this is at the heart of any personalised energy strategy; and the answers are not obvious at first glance. Indeed, the true benefits of autonomous PV have been widely misinterpreted in the race to market, and the subsequent mismanagement of expectations has led to something of a public image problem, in the industry and in the world at large.

In this chapter we visit some applications of 'off-grid solar' and see what it has to offer. We see how lowest-cost comparisons are misguided and restrictive to the stand-alone PV format. We also investigate how these and other culprits contribute to the modern-day bad reputation it suffers.

The wider impact is on those who influence the developing world energy sector. They are seeing bad examples of a good technology, reinforced by a new generation of bad claims and low-quality products and solutions. This reputation must be addressed by our knowledge-sharing strategy.

The widespread poor reputation of 'solar' poses by far the greatest challenge to the whole notion of 'Selling Daylight' – a business strategy with a potential identity problem. How did this come about?

Before we bring together the hardware and knowledge-sharing methodologies of the last two chapters, then, let us first visit some of the applications of 'off-grid solar' and see what it has to offer. We will see how lowest-cost comparisons are misguided and restrictive to the stand-alone PV format. We will also investigate how these and other culprits contribute to the modern-day bad reputation it suffers. This reputation must be addressed by our knowledge-sharing strategy.

The origins of off-grid PV's reputation

The original value propositions for the earliest autonomous PV systems have remained relevant for over three decades. Of course, the world is a very different place since the 1980s, particularly in terms of international communications, logistics and advances in technology. Solid-state devices have massively improved in reliability and efficiency, and this has drastically reduced the

http://dx.doi.org/10.3362/9781788530705.008

cost, size and complexity of PV systems. Nevertheless, it pays to have a general understanding of projects from the 1980s and 1990s if we are to appreciate their impact on local societies and the wider donor sector. Their value propositions are applicable to many areas of the PV sector.

In the beginning, all PV systems were autonomous. In 1953/4 the first PV cells were developed in Bell Laboratories in the United States – greeted far and wide with giddy post-war optimism of what this new technology could achieve in the near future. *The New York Times* declared the invention 'the beginning of a new era, leading eventually to the realization of one of mankind's most cherished dreams – the harnessing of the almost limitless energy of the sun for the uses of civilization' (The American Physical Society, 2009).

The first satellite with PV cells, Vanguard 1, was deployed in 1958. This was a fundamental endorsement of the technology but also, *crucially, the value principle*. No one talked about the cost of the PV system; instead they talked about what PV had made possible. People within the industry and the popular media praised the value of PV for the communication system it had enabled.

Back on Earth during the years following Vanguard, PV became successful for digital watches, calculators and a range of consumer products (the Edmund Scientific Catalogue included PV from 1975). The 1960s also saw the first terrestrial systems, and with them the first strong commercial demonstration of PV (notably being used for lighthouses in Japan). But it was the 1970s when PV truly showed its earthly potential. It was telecommunications infrastructure that epitomised commercial value, somewhat ironically in the Middle East, the extreme sunlight resource ideal for a relatively modest PV array.

The cost of using fuel-based generators or extending the grid was prohibitive, and in many cases the terrain made them impossible. Under these circumstances, PV systems using 1970s technology – PV, batteries and electrical switchgear – and 20 times as expensive as are today, were delivered by helicopter and were still justified commercially. The extreme costs were acceptable because of the critical need for the service that the PV would enable. This commercial viability was a precursor for what was now possible.

There had been various smaller-scale demonstrations of PV before these large bespoke systems started to be used. I have images of PV-powered telephone boxes from the mid-1970s and various monitoring devices used in very remote areas, incorporating small amounts of shiny grey PV. (These grey cells pre-date anti-reflective coatings. Modern cells are almost black. Cells that are blue are older technology.) A handful of small solar companies, mostly run by engineers and scientists, started to produce and sell PV solutions for applications such as navigation buoys, radio communications, lighting, monitoring systems, safety shut-off devices (including for mains electricity systems), flow valve control and water pumping. But the larger systems were critical for the PV industry to become established. The most valuable of these were not widely known, mostly because they were commercially sensitive or for confidential military applications.

Bespoke systems

All large autonomous PV systems were, and mostly still are, designed specifically for the requirements of a given application. Systems are sized for a single location, and tailored to the specific incident sunlight resource as well as the local environment. Anything that could reduce the actual light energy reaching the PV is taken into account (such as the orientation of the panels, shading from trees, dust, bird mess, etc.). The systems are electrically configured according to what is being powered (telecoms equipment is usually 48V DC for example). Safety and security are also considered specifically with regard to the location and the needs of the customer.

The systems are difficult to design, and their requirements stringent. In nearly all cases the systems need to have long operational lives, with low maintenance. Telecommunications was an important application for PV from the late 1970s and throughout the 1980s, primarily because its infrastructure demands longevity and reliable operation as a matter of commercial necessity.

A solar platform for oil and gas

But it was a different sector that truly utilised PV's value propositions. Again, it is somewhat ironic that one of the major enablers of the early terrestrial off-grid PV sector was the oil and gas industry. Unmanned offshore oil and gas platforms have employed autonomous PV systems since the 1970s. These metal structures are basically the interface between the vertical pipe coming up from the oil or gas reserve, and the horizontal 'surface' pipe carrying the product to a collection location.

These platforms need constant monitoring; control and safety systems ensure that the valuable commodity keeps flowing, or is limited or stopped in an efficient and safe manner. They are regularly powered by PV because they are designed for long operational life (30 years or more) in remote locations; they represent significant cost if they are not functioning and significant safety implications if they fail. The explosive gases that accompany oil or gas extraction often preclude the use of high voltage switchgear (mains electricity) or the spark and heat hazards of, say, diesel generators.

Cathodic protection and SCADA

Two further applications for off-grid solar emerged within the fossil fuel sector from the 1970s onwards. *Cathodic protection* is a way to prevent the corrosion of metalwork. The primary applications are for pipelines but they can be used for other large metal structures, such as bridges and water tanks. The basic principle is to use a 'sacrificial' anode (a carefully selected type of metal) to draw corrosion away from the metal you wish to protect (called the cathode). For large structures, a voltage and current are applied to the anode and cathode to increase the effect of the protection. Here, stand-alone PV ensures or extends the lifetime of the expensive pipes that carry corrosive fossil fuels. Industrial PV power systems for cathodic protection are designed for 20 to 30 years; they demonstrate the long-term effectiveness of the technology, when designed correctly.

At this point, contemplate for a moment that it was the fossil fuel industry which first enabled the terrestrial PV industry. Stand-alone PV continues to enable a significant proportion of fossil fuel production. Approximately one-third of all fossil fuels are used to produce grid electricity, and the vast majority of the PV industry is presently focused on grid-connected systems which depend on fossil fuels to keep that grid operational.

SCADA (*Supervisory Control and Data Acquisition*) generally refers to an industrial computer system that monitors and controls processes; for example, safety systems such as shutdown and pressure release valves. SCADA systems are typically used to control geographically dispersed assets (such as the oil or gas platforms and pipelines described above); monitoring and transmitting flow, pressure and other data with great accuracy.

Being safety systems, SCADA demands absolutely guaranteed dependability in the power supply. Multiple generation systems are combined – such as diesel generators and/or mains electricity (though often neither are commercially or logistically possible) with battery backup. Alternatively, SCADA uses autonomous PV, subject to detailed design requirements and multiple layers of redundancy. These are another perfect demonstration that PV-enabled energy systems can be dependable and reliable over many decades of use; once again, if they are designed correctly.

Design effort and cost

The wellhead, pipeline, SCADA and telecoms applications are all designed to operate for decades in far-flung places, and where downtime is expensive. The value propositions of bespoke stand-alone PV marry perfectly with these requirements, which is why the high initial cost is entirely justified. These stand-alone systems are not trivial to design and they are expensive to supply, with significant design work required before even a ballpark quote can be made to the customer. The need for dependability means that the requirements of the customers are very detailed, and wrapped in multiple layers of legal conditions and obligations. But you can't have one without the other. Without the specialised efforts of PV system designers spending weeks or months developing these solutions, you cannot expect to have a 30-year service life with very little maintenance overhead. Unfortunately, this is exactly what many have tried to do.

> PV-enabled energy systems can be dependable and reliable
> over many decades of use if they are designed correctly

The respectability of industrial off-grid systems spread beyond the PV industry and those of its specialist customers from the early 1980s: they are expensive initially, but they are reliable and require very little maintenance; they are silent, have no emissions and can power a wide range of electrical equipment in very remote and challenging locations.

So, why not use them for people without mains electricity? Why not use them for 'rural electrification' and help those in or near poverty?

Mini-grids

The non-industrial versions of bespoke systems were generally referred to as mini-grids; and later, micro-grids. The names are our first indicator of why they didn't work: they tried to take the place of what the grid does. As we have seen, the expectation of a grid (implicit or explicit) is the provision of boundless electricity without ever having to think too much about it.

A number of large mini-grid projects were implemented in the mid-to-late 1980s, funded by the World Bank and other organisations. The concept was simple: put a large autonomous PV system in the centre of a village and run cables to each household. On paper, the economies of scale make sense, because one big PV system is cheaper than lots of smaller ones. But nearly all of these projects failed. The systems were unreliable, served fewer people than predicted, were more expensive to maintain or lasted far shorter than intended by their design.

Primarily, though, there was a lack of 'ownership' by the users. As with so many so-called aid projects, the equipment was installed and commissioned; then the donor organisation moved on. Daily activity around the PV made it dirty, and cost constraints ruled out the use of many environmental protection measures applied to industrial systems. This called for some basic infrequent maintenance, such as cleaning PV panels, checking electrical connections (especially to the battery and control system) and assessing the conditions of the batteries. The maintenance skills required grew in proportion with the systems. (Checking electrical connections on a hand-held PV module and battery can be performed by pretty much anyone with a little training. Checking them on a large system with enough potential power to kill you is another matter entirely.) So systems were not maintained; they steadily underperformed. And then failed.

The problem lay in the cost analysis used to size and implement these projects. The reasoning, broadly, was this: x number of households require y amounts of energy each, so (with an estimation of sunlight) the mini-grid PV system must be z capacity. And the systems were 'designed' to optimise dollars spent per household reached. Central to the cost argument was that electricity from the mini-grid needed to be cheaper than what households were paying for kerosene or diesel (or even batteries).

This type of financial analysis always puts pressure on reliability. When, inevitably, consumption per household exceeds predictions then more energy is drawn daily from the PV system battery than it was designed for. This drastically reduces both the battery's usable capacity and lifespan.

Users were not sufficiently skilled to maintain these large systems, but neither were they aware of how they operated, nor how their personal consumption would affect the overall value. Energy-efficient appliances were not readily available in the 1980s; so each act of inefficiency by the user had a more dramatic effect than they would on a more modern system. The cumulative effect of leaving a 50-watt light bulb on overnight is far more damaging than it is for a 5-watt LED version (of the same or better brightness).

The value of these installations for rural communities is undeniable, and they are often cited to justify new systems in other regions. But while there are some shining examples, the majority of applications have not performed effectively.

Consequently, mini-grids are a major contributor to the ill repute of PV in the developing world. Despite regular reports since the early 1990s highlighting the flaws of these systems, they continued to be installed; each new group looking to 'help' with or 'address' rural electrification thinking they won't make the same mistakes as in the past.

For these systems to even have a chance of serving their customers effectively, certain basic requirements must be met.

- Clear ownership of the hardware (both commercially and locally);
- Trained local personnel to operate and maintain the system;
- A tangible cost involved to the user, to engender a sense of value in the service;
- A local organisation to collect revenues and pay for maintenance;
- Established suppliers for replacement parts (particularly the battery; but including any other component);
- Initial and regular training of everyone concerned with the system; identifying and employing a skilled training force.

Each new iteration of the mini-grid format has begun to address more of these issues; but as a generic format they have still failed to deliver long-term, cost-effective or reliable energy services.

A new interest in mini-grids

Over the last few years a new interest in mini-grids has developed. It started within PV industry press and development circles, often rebranded as 'micro-grids', and has now reached policy makers and commercial developers. What sets these 'new' systems apart from those of the past are aspects such as smart meters for each household, limitations on household consumption, pay-as-you-go systems, better education and training, and, as ever, lower cost technologies (particularly PV modules). There is also a renewed push for rural electrification, due to the demand for communications and online access.

Implementing any new system carries a responsibility – it must unequivocally help reverse, rather than reinforce, PV's troubled legacy

Mini-grids appear to be the best solution for agencies concerned with education, healthcare and social welfare. On paper this is understandable. Since the problems of the past, user devices have improved in terms of efficiency. Addressing modern-day energy needs in this way reduces the scale and complexity of these energy systems, and allows them to serve higher numbers of people. As a result, the value *potential* is therefore much higher than in previous generations.

Mini-grids have certainly improved enormously in their value; they represent significant potential for providing energy services in the developing world. But implementing any new system carries a responsibility – it needs to be exceptional in its performance and must unequivocally help reverse, rather than reinforce, PV's troubled legacy.

Comparing industrial and public applications

The frustration for autonomous energy specialists such as myself is that, for the last 30 years, industrial bespoke applications have remained a paragon of long-term value; while systems of the same physical size (sometimes using the same core components) have repeatedly undermined the more public side of the technology format.

On one hand, oil, gas and telecoms infrastructures have continued to enjoy the benefits of bespoke stand-alone systems, while rural communities, hospitals and schools have repeatedly been let down by poor implementation, or shied from the opportunity of PV technology due to its public standing. The same is true of smaller systems in the industrial realm versus their counterparts in the commercial and retail sectors.

The buoyant value of PV

The value of autonomous PV does not necessarily stem from making batteries last as long as possible. Instead the aim is to make the batteries last long *enough* so that they can be replaced during scheduled maintenance visits. The design must ensure that the PV energy supply is not the weakest link in the overall installation.

Take the example of a navigation buoy floating in the sea. Buoys are tethered around harbours, busy shipping lanes and dangerous areas. At night, or in low visibility, a light flashes to warn boats of potential danger. If you power one of these lights with a battery on its own, it may last around four weeks before it needs replacing or recharging. (When PV was first applied as an energy source in the late 1970s, before LEDs were available, the buoys used high intensity, high power light bulbs.)

A longer lasting battery requires more capacity, meaning an increase in size and weight – crucial to us because it is installed into a floating unit. There is a natural limit to how big the battery can be (the weight that two people can lift from a boat, for example) and therefore how long it will last powering that particular device. By adding PV to the buoy the battery receives charge every day, and it now becomes technically possible to make that battery last 5 years.

This giant improvement would represent a value proposition in itself. But the practical requirement is more subtle – it is time consuming, logistically challenging and expensive to inspect and maintain navigation buoys. But it is necessary to check them at least annually, for corrosion, physical damage and the security of its components. And while you are there you can change the battery. The maintenance schedule provides our *critical path* for determining the PV requirement.

The battery is not a high-cost item – typically around a hundred dollars. So as long as the PV power system will comfortably last a year, you can justify a system like this on the offset costs of *not having to visit the buoy every month* just to change the battery. You would design more than a year's battery life into a critical system like this, to account for damage or unusually bad weather – which reduces solar charging while increasing the demands of the buoy light. But the annual maintenance schedule will keep the likelihood of failure (and costly interim technical visits) very low.

From industrial excellence to the public domain

Bespoke stand-alone systems established huge potential for rural electrification in the developing world. Sadly, this has never yet been realised. Once established applications set imaginations alive (such as the buoys described above), the higher profile commercial applications failed to deliver.

Street lighting is an obvious use of PV, and attempts to implement solutions go back at least 15 years. The value proposition is simple: standard street lights require a mains electrical cable to be installed, but this is prohibitively expensive for rural areas, whereas autonomous PV can simply be installed anywhere without infrastructure or logistics issues (think of the parking meters described earlier). Unfortunately, the high-profile nature of street lighting meant that poor implementation was visible to everyone, and the refrain 'Solar street lights don't work' became accepted as immutable fact.

The same problems are played out: an obsession with initial cost, while trying to replicate the standard models. Mini-grids attempted to replicate the national grids and encountered the same complacency it induces in users. Lighting projects tried to replicate the output of High Intensity Discharge (HID) lamps with lower energy consumption technologies, in order to keep the cost of the PV system down.

This is a shame. Today there are effective street lighting products which use high-intensity LED lighting while retaining energy efficiency.

Careful design techniques, such as efficient optics and reflector arrangements (as well as simpler changes, such as placing the lights closer to the ground) have proven highly effective. But the familiar public relations problem, combined with a short-sighted cost motive, continues to hinder any widespread deployment of these newer, strong performing products.

Instead we see the lowest cost products fail to perform and ruin the sector for everyone.

Frustratingly, street lighting applications are often seen as a way of demonstrating a commitment to 'green' technology or carbon reduction. They are often installed in prominent locations such as the main roadways between airports and their host cities as a showcase for progressive or green credentials. Kampala, Addis Ababa and Nairobi sport hundreds of solar streetlights, and last time I dared to look, most were not functioning. The cheap ones look ugly, don't work, are hard to maintain, are badly designed and were a waste of money. That money would have taken months to justify, and instead of promoting the value of PV, it resulted in the exact opposite.

Groundwater pumping

We cannot talk of the early days of 'off-grid solar' without mentioning groundwater pumping, one of its original applications. Although pumping systems were used for industrial applications, water pumping – like mini-grids and street lights – was largely developed in more public environments. Some notoriety emerged here too: mostly resulting from taking a difficult environment for electric equipment, and failing to apply sufficiently strong designs to the challenge. Moving parts, water and electricity do not play well together. Add heat, humidity and the damage of equipment in transit, and it is no surprise that the early systems encountered so many failures. A number of outbreaks of disease followed.

Water is essential to life, so the value proposition here is obvious and simple: water is pumped from underground or from a local source for irrigation, for livestock and wildlife, for storage and for people. The provision of water is one of the simplest solar value concepts to understand. Water pumping is one of the few PV applications that requires no batteries (though they can be used to strategic effect). The needs of a given location can be met by pumping groundwater up into a storage vessel only during sunlight hours.

I am pleased to say that 'solar water pumping', following many years of growing pains, has become a mature sector. Professional companies are producing highly dependable, high-efficiency DC water pumps. Many also supply tailored PV energy systems and provide long guarantees for their complete solutions. And because the energy requirements of the pumps have reduced so much since the early days, stand-alone PV is easily justified on an economic basis alone.

The provision of water is one of the simplest
solar value concepts to understand

However, it is not all roses. Many organisations engaged in water provision do not adequately understand PV or electrical pumping (or water, for that matter). AC water pumps are regularly employed despite the widespread availability of less energy-demanding DC pumps. They are therefore missing a clear opportunity for smaller, cheaper and more reliable PV solutions.

It would be amiss not to make reference to water quality. I have come to appreciate there is no such thing as common sense when it comes to people trying to 'help'. Water should not be moved through corrodible materials and must only be stored in specially designed containers, properly sealed from the environment; failure in this regard risks danger to the public. Where this has occurred (unfortunately it is not uncommon), PV becomes as tainted as the polluted water. And the notion that 'Solar water pumping is dangerous to health...' unfairly enters the public imagination.

Autonomous PV applications in industrialised societies

An improved understanding of stand-alone PV potential grew through the 1980s and 1990s, both with suppliers and potential users, and with it grew the number of applications. Some had obvious value; some were a little obscure. Some were executed very well, and others let the customers down. The success stories among the industrial applications, such as monitoring devices, are generally out of public sight. Instead, the most prominent examples of stand-alone PV, as we shall see below, are doing little to help or develop the generic technology.

Calculators
Calculators used to be the most prominent and recognisable PV application. They have now largely been superseded by calculator functions on smartphones and computers.

In the industrialised world, there are many perfectly reliable and valuable industrial applications, such as monitoring devices. However, these are generally out of public sight

---

Garden 'lights'
Now one of the most visible applications, these entered the spiral of lowest cost competition many years ago. They are now so cheap and nasty that they usually have to be replaced each spring, or just thrown away even before the end of summer. They do not really qualify as 'lights' so much as little glowing things that illuminate rather than produce useful light. They are sold as cheap gimmicks and with little guidance on how to use them. They therefore end up behind trees and in back gardens, shaded by houses or sheds, and broken by pets, or even harsh wind.

'Flexible' solar
So-called flexible solar panels really are pretty cool in what they promise: a lightweight, foldable or otherwise versatile sheet of tech that will simply charge your personal devices when you need it to. But the reality is far from this – especially as some of the 'flexible' modules are actually quite brittle. They break regularly with mild physical stress and are sold at massive mark-up.

The technology might cost, say, $5 for a 1-watt panel, but is sold for as much as $100 (this is a real example at the time of writing).

Solar bags
A popular phase of solar bags and backpacks has largely passed due to several issues. Firstly, the 'solar' is vulnerable – either the PV panels have glass or they are the so-called flexible versions and not very rugged. They are usually not supplied with electronics or the correct battery (if one is included at all) so they are not very effective. Most of all, being retail products subject to the huge margins of the retail system, they are expensive: the bag itself might cost $10 to produce with an extra $5 of PV on it, yet the bag as a retail product is sold for $300 (again, real figures). Performance is unpredictable because the bag's orientation is constantly changing with respect to the sun. The most natural orientation of the bag and its PV actually requires the sun to be low in the sky, which is much weaker in terms of incident energy. (The impractical alternative is to leave your bag outside, facing the sun for hours at a time...) Similar offenders are house number illuminations and movement sensor security lights, whose effectiveness is limited unless they are installed in very sunny locations; not to mention gimmicky solar toys designed for indoor use.

DIY kits
Stores such as RadioShack in the USA sell solar PV kits; in theory, they enjoy just the right type of tech-friendly customer base to engage with the concept. Frustratingly, as with the bags, the user equipment is typically not sold with the PV (nor is the battery in many cases) so the system is inefficient at the outset, no matter how much attention is applied to its use. All that hard-earned energy generated by the PV is mostly wasted.

Parking meters
Prominent and visible, these solar parking meters initially did wonders for the promotion of stand-alone PV. Unfortunately, some members of the public have chosen to disable them by shading them with dirt or black plastic bags, in order to claim exemption from buying a ticket. This vulnerability to public behaviour is used as a further example to undermine the credibility of off-grid systems and generic PV.

Performance claims
Fuelling the negative exposure for PV is the practice of selling products into inappropriate regions. Performance claims on packaging are usually the best-case scenario, advertising the best possible (or purely theoretical) performance, without an adequate disclaimer of the expected real-world capability. Products often advertise their performance specified for somewhere like the sunny South of France, or southern California; and instead of producing alternative versions, these products are sold into far less attractive solar locations. The products are never going to function properly in Scotland or Seattle.

The legacy of PV technology

Since the turn of the millennium, the PV industry has concerned itself almost exclusively with grid-connected applications. Some have cited this as the cause of the off-grid PV sector decline. But, in truth, its central concepts are not in question; rather it is the inappropriate or ineffective *application* of the generic stand-alone PV format which has steadily undermined its real and its perceived value. Off-grid PV has a proven track record of high value for industrial systems; but a reputation for underperforming everywhere else. The reputational damage from failed PV systems of the 1980s is still being felt to this day. Modern examples, primarily in portable and consumer goods, are doing nothing to build confidence in the technology. After the sheer time that has passed, it is vital to appreciate how this poor reputation arose, and the economic culture which gave rise to it. The reputation of PV is the biggest challenge for this whole strategy, and the scale of the problem must be appreciated if it is ever to be overcome.

The impact of poor products goes much further than
a fall in sales. The wider impact is on those who
influence the developing world energy sector

In the richer nations, the impact of poor products goes much further than a fall in direct sales. After all, a strong value proposition only exists for niche applications *away* from mains electricity, so there is a knock-on effect, which impacts those who influence the *developing world* energy sector. Energy-poor nations crave reliability in their energy services. Stand-alone PV can address the majority of the challenges, yet it is overlooked: seen as a supporting technology; a small piece in a big energy supply puzzle. All the while, other, costlier, more polluting and lower-value energy solutions are implemented to try to meet the immediate demand. This is hardly surprising in regions where stories of unfulfilled promises linger for generations. The decision makers, then, are seeing bad examples of a good technology, reinforced by a new generation of bad claims and low-quality products and solutions.

PART THREE

Major elements of the Selling Daylight strategy

CHAPTER 9
Focusing on quality and value

Abstract

Despite the efforts of well-intentioned organisations, sub-standard stand-alone PV products contribute to keeping people in poverty, all the while selling short the most capable technologies for alleviating it.

In this chapter we discuss what 'value' and 'quality' mean in the context of energy services for the developing world.

We explain why there has to be a primary economic justification, and that a PV energy service's value disappears below a certain threshold, beneath which they hinder rather than help development.

In developing nations, energy is consistently poor in quality. In a blinkered market where purchase cost is the dominant metric, quality becomes compromised. And when quality suffers – be it in the design, components, usability, operational costs, service life or environmental impact – its resulting value plummets.

Quality must not only ensure actual value; it must help make value easier to understand. The responsibility of shrewd marketing cannot be downplayed here.

The state of play, then, is this. The early promise of industrial stand-alone PV systems has failed to transfer into the public domain. Modern-day portable and consumer goods – trendy, expensive and ineffectual – have sustained a bad reputation for all things 'solar'. Meanwhile, in the developing world, quality solutions with genuine value propositions are giving way to constant cost competition. But what do we mean by 'quality' or 'value'? What opportunities are we throwing away in the pursuit of a competitive sale price?

Developing nations and off-grid PV

At the time of writing, the highest profile PV sectors in the developing world are mobile phone charging, lighting and solar home systems. What are these doing to promote the future of the underlying generic technology?

Lights and chargers

Around 2005, I bought a few solar mobile phone chargers in Africa and Asia when they first started hitting the local stores. None of them functioned reliably, regardless of how much sun they received. The first products to reach the market had no energy storage, so the phone had to be left out in the sun for

http://dx.doi.org/10.3362/9781788530705.009

charging; not an auspicious start. They had inadequate, poor quality PV, and many damaged the phones they were meant to be charging.

Their quality has remained appallingly low. The majority now include an internal battery. But these are typically undersized and of low quality. The only half-decent products are sold as multi-purpose chargers, boasting USB sockets and sometimes mains charging capability. Nowadays the most credible phone chargers are often lighting devices which additionally charge phones.

Since around 2008, there has been a massive uptake of solar lighting products; partly due to significant improvements in LED and battery technologies, and from large market development initiatives such as the World Bank's Lighting Africa programme (which later became Lighting Global). There are now hundreds of small solar-powered LED devices including torches, reading lights, fishing lanterns and products for craftwork and food preparation. Traditionally the most common solar lights were called 'solar lanterns' after the kerosene lamps they sought to replace. This is less common now, partly due to the huge appetite for mobile phone charging – which many devices are now capable of providing. The growth of this market, particularly for multi-purpose products, has engendered a new name: 'pico solar' (noticeably, not pico 'PV'.)

Here is a straightforward enough proposition: to offset the cost of kerosene, candles and disposable batteries. There is a sound financial justification (as long as the product has a reasonable operational life), supported by clear health, safety and environmental benefits. This market growth is welcome, but it has resulted in a deluge of low quality generic lighting products. These are now a commodity item. There remain companies with quality, ethically focused products, but a growing majority are traded (with little to no user engagement) alongside cheap mobile phones, radios, disposable batteries and other tech consumables.

You can buy a small solar torch or 'reading' light for just a few US dollars. It will not be worth the money. Fundamentally, any light that provides one hour of operation, or is bright enough to read by only when held against the page (despite the best efforts of the World Bank, there remains no standard requirement for lighting), or whose battery fails after two months, or stops working when dropped from waist height, is evidently failing to provide practical value to the user. I have been testing, using and designing pico lights for the last 12 years. I estimate that easily 90% of the products on the market fail in this regard; sometimes even falling short of kerosene lamps or disposable batteries in their capability. Meanwhile, a product which is fit for purpose and meets the users' basic requirements will almost certainly not be the cheapest; indeed, depending on the supply chain and location, it could be $30 or more. But sometimes it is possible to provide significant value for just a few dollars above the base-level product, by way of rugged housings, waterproof construction, remote control and other features.

It is often difficult for the untrained eye to distinguish the good from the bad. Neither suppliers nor customers know the value of more expensive, well designed products over their cheaper, unreliable counterparts. Helping to understand a product's true value, determined from its performance and features before its purchase cost, is a genuine responsibility, which can be supported in part by our Visual Instruction System.

During 2014 to 2016 I asked knowledgeable friends within the developing world lighting sector the same two questions. How many pico solar lights have been sold in the last 10 years? And how many are still *working* – actually offsetting kerosene, candles and batteries? This is an inexact science, with many assumptions. But consider the *scale* of the discrepancy in the answers I received: approximately 50 million and 5 million, respectively.

Solar home systems (SHS)

When mini-grid PV first began to draw a negative press, the general consensus was to take the systems closer to the user: instead of a single large system powering 50 houses, why not put a small system on each house? This migration away from the larger mini-grid format helped establish the generic term 'solar home system (SHS)'. (Note again, not 'PV' home system.) While personal and small-scale lighting represents the biggest sector by far for autonomous PV in the developing world, solar home systems are catching up quickly.

They consist of a central unit, housing a battery and some control electronics. They are supplied with a PV panel and cables for connecting to various devices. Most products include two to four lights, but are often capable of powering more. As the name suggests, these systems are primarily for domestic applications, such as mobile phone charging, fans, radio and TV. As with the small pico lights the greater demand has led to more suppliers wanting a share of the market. The result is a wide range of products, whose quality differs wildly.

I have worked with 40 or so different systems, loosely within the SHS category. Reassuringly, most perform the task of lighting fairly well; but they fall short of enabling services such as refrigeration in terms of operational hours and reliability. These systems are close to being a consumer product format, so the preoccupation with initial cost has already taken hold. As with the solar lights, there are some very cheap products out there, which are doing nothing to help reputations. A good, reliable system, with useful features (such as a user display), rugged construction and general fitness for purpose is likely to be over $100, but can be twice that depending on location. Selecting the correct product will involve understanding the needs, the specifics of the environment, and only then, lastly, looking at cost options. Few customers of consumer products are taught how to do this.

Helping and hindering the poor

The poorest people of the world badly need electrical services, but they do not have the commercial leverage to attract quality solutions. Well-intentioned individuals and groups seek to help the poor directly. This conceals a fundamental problem, because it reinforces the perceived need for low-cost, low-end products. I want to make it clear that below a certain level of quality – that is, of *minimal value to the user* – stand-alone PV is simply not worth having. Put another way, below a certain price point, PV products are of no benefit; particularly to a developing world user for whom such a device is not a leisure item but a life-changer. Instead, sub-standard products contribute to keeping people in poverty, all the while selling short the most capable technologies for alleviating it.

It is honourable to want to make products more accessible to the poor and to try to reduce the cost of energy services. But as soon as you demonstrate sales of any scale, the opportunists, who are not operating from a base of well meaning, will make a product cheaper than the competition, or otherwise deceive the customer with false claims of product performance. Where the financial profit motive crosses into unscrupulous behaviour, quality and value are sacrificed for the sake of a margin. In a market that operates primarily on lowest cost, this puts pressure on the philanthropists, either to spend money on marketing and staff to overcome direct cost comparisons, or to compete on cost. The result is commonly a bit of both.

Sub-standard products contribute to keeping people in poverty, all the while selling short the most capable technologies for alleviating it

---

The quality and value threshold

When designed correctly, in adherence to a strong central value proposition, autonomous PV solutions can address a huge number of energy service requirements in the developing world. Industrial systems continue to prove this. However, a PV energy service's value disappears below a certain threshold, beneath which they hinder rather than help development. That threshold is defined by the product design, the components and the user's ability to operate the device itself.

Take the example of a car battery. You expect it to start the car, be charged by the car and power all the electrical equipment within the car. You expect it to do this for a 'reasonable' period of time – say, an absolute minimum of a year, but far more likely 2 to 5 years. The battery has a direct impact on the functioning value of the car and the quality of service it can deliver.

Say we have low-cost car batteries being readily available on the market. But we can only start the car for the first few weeks before struggling. Or, having got going, no electrical equipment can be used (including the headlights). Or again, we can start the car, the equipment is operable, but the battery needs to be replaced every month. Clearly we're not going to get very far.

A PV energy service's value disappears below a certain threshold, beneath which they hinder rather than help development

---

Now consider a stand-alone PV system which, after all, is a battery charger. Above our quality threshold, autonomous PV continues improving; the higher the quality, the higher the value. Below this threshold, value is simply absent.

It is not a case of 'some value' or 'limited value' or 'better than nothing'. There is no such thing as a low-value PV system: they have value or they do not.

Economic value

There are numerous advantages in being enabled or supported by PV, but first there has to be a primary economic justification. As examples, the use of PV must deliver the following:

- Enable a service for which there is no realistic alternative;
- Save the user money over the operational life of the product;
- Make a service more accessible or convenient to the user, or less problematic to the supplier.

There are many secondary areas of value, such as carbon offsetting, demonstrating a visible environmental statement, or simply being a 'cool' product. But these have proven ineffective when it comes to the widespread uptake of stand-alone PV.

This is true in the examples we have seen already. In many of the larger PV applications such as the telecoms repeater stations in Oman, there is very little secondary value from being 'green'. The value lies in having a reliable remote power supply with a low maintenance requirement – essential for an installation on top of a mountain.

On our streets, parking meters have an obvious direct economic value to the supplier, while for bus shelters and bus stop lighting, the value may be different. Transport for London Ltd (guess what they do) installed PV lighting systems on their bus shelters and bus stops as part of a wider initiative within London to reduce emissions and demonstrate commitment to the environment; but this was not a sufficient primary financial justification. More pertinent was that the bus shelters and stops were moved regularly as bus routes were adjusted in line with traffic flow developments and other changes. Furthermore, interestingly, bus shelters are hit by buses on a worryingly regular basis, and the expense of moving or repairing mains electricity connections each time is offset by the PV solution.

Our calculator example from earlier also illustrates this. There is no need to replace a battery; its design is more attractive without an accessible battery compartment; and it can be seen as a 'green' product. But these are secondary to its primary value proposition – that PV combined with a small rechargeable battery is cheaper and longer-lasting than a conventional battery on its own.

For all of these applications the secondary benefits help make the case, but they do not justify the use of PV on their own. It doesn't matter how sexy the product looks or how much hype is generated around it: if it doesn't offer real value, ultimately the trimmings won't help.

So many manufacturers of consumer PV products fail to address this central, intrinsic value. Think of the portable solar chargers we described in industrialised regions. They range from 10 dollars to several hundred, and differ in quality, but the fundamental problem is that consumers in the industrialised world simply do not need the PV; they are never far from the mains grid or a vehicle if they wish to charge their devices. Here, value exists only in niche markets such as camping, and many such products compromise the cost or size of the PV itself, so it never typically receives the sunlight needed to generate useful energy. Its value proposition has been undermined, in other words, and no amount of cool branding or hipster cachet can bring it back.

In search of quality

Energy in the developing world needs to be dependable. For our initial target market – those with money but unreliable grid electricity – this is the requirement which holds the most obvious value. In order to enable quality services, the electricity supply needs to be considered as dependable as the mains grid in industrialised nations. If this is not possible, it has to be at least *predictable*, or better still, *controllable*, so that users can prioritise and schedule the services they need. In the developing world, stand-alone PV can be more dependable than any other electricity supply. It can also be predictable and controllable by the user.

So far, in order to make sense of abstract or subjective language, we are using the word 'value' in the sense of a 'value proposition'; to mean the principal, tangible, compelling reason for choosing our energy services. Similarly, can we pin down a worldly definition of 'quality' – one which gives us a realistic means of judging products and services? I believe we can.

Quality means fitness for purpose. It does not mean expensive. A Rolls-Royce luxury car is not a quality commuter vehicle in a congested city with limited parking spaces. A solid gold tennis racquet is not a quality product where lightness and flexibility are essential. A quality solution is the best fit for the multiple needs of its users. High cost does not mean high quality; and although high quality enables high value, it does not guarantee it. Most importantly of all, low cost does not mean good value. A $100 million coal-fuelled power station – commissioned because it has the lowest cost per kWh of the available generator technologies – is not a quality solution in a landlocked country, with no coal of its own and a widely distributed population.

In developing nations, energy is consistently poor in quality. Central power stations are by definition unsuitable for these often vast or mountainous land masses; and distributed solutions fail to meet actual needs, lacking resources, maintenance, fuel and financial support. Yet we keep seeing the same dogged approach to energy provision, despite 40 years of attempts to provide 'rural electrification', 'electricity as a basic human right' or 'energy access for all', which mostly fail or under-perform.

Quality means 'fitness for purpose'

Some examples? A diesel generator is installed to provide for 10 light bulbs. The generator uses the same amount of fuel for those 10 lights as it would for 50; it requires skilled maintenance and constant refuelling. Elsewhere, a mains grid connection with massive capacity is installed to a building where only a few computers are required. Meanwhile, for some large systems, PV is typically procured as a power supply, or even worse, as a set of components. This is analogous to specifying a car primarily on how fast it can travel without adequately emphasising the importance of its features, such as fuel efficiency, safety or comfort. Buying autonomous PV as a power supply prevents professional designers using their skills to optimise the quality of the overall solution.

All of which is a shame; because when designed and implemented according to requirement, stand-alone PV is the right solution for energy-poor countries because it is a *quality* solution, in the context we have described above. It means dependable (or at least predictable) energy services, where they are needed, that do what the user needs without including unnecessary complexity, with a cost that does the same: you pay for what you need and not for what you don't need. This may sound blindingly obvious, but it is rarely practised. For stand-alone PV, the overall value is heavily influenced by the specifics of what is being powered; who will use the equipment, where and for what primary purpose.

The myth of purchase cost

Why do so many solutions end up unfit for purpose? One clue is the widespread obsession with *how* energy is provided, rather than *what* energy can enable. Outwardly, the public dialogue concerns education, health, communication, entertainment or light; but when it comes to purchasing, we see a tunnel vision focus on the unit cost of the actual computers, TV, fridges, cheap phone charging and internet access; followed by cheap power supplies for all of these.

In a blinkered market where purchase cost is the dominant metric, quality becomes compromised. And when quality suffers – be it in the design, components, usability, operational costs, service life or environmental impact – its resulting value plummets. Such short-termism consumes precious personal and national funds, preventing the implementation of energy infrastructure as a reliable service enabler. This continuing widespread lack of understanding remains an obstacle which we will need to overcome in our search for quality.

> In a blinkered market where purchase cost is the
> dominant metric, quality is always compromised

The cost of cheapness

It is not just autonomous energy services that suffer in this way. All low-cost options, whether they are cheap televisions, fridges or other appliances, nearly always consume more electricity and experience shorter operational lives than their higher cost counterparts. Many, such as glass door fridges, place an unnecessarily high energy burden on limited capacity mains distribution networks.

The cost of electricity is often overlooked. Even a small capacity, cheap fridge may cost as much as $300 per year to run, whereas the low energy version could cost as little as $30. Indeed, the privilege of having mains electricity is slowly being recognised as a burden for some due to the high expenditure – which cannot simply be removed during times of hardship. But the alternatives are not available to them.

We have said that people who use these services tend not to know the difference between good quality and poorly made products. Presently, they will need to. PV suffers these issues also because cheap appliances equate to costly PV energy supplies. As cost increases, the value it creates must be more compelling and obvious, if we are to reach customers in the numbers needed. Quality must not only ensure actual value; it must help make value easier to understand. The responsibility of shrewd marketing cannot be downplayed here.

Elements of quality and value

I have expended a great deal of words on the deficiencies of PV in the market, the problems we face in terms of reaching users and stakeholders, and the struggle against a widespread poor reputation. Now, having asserted real-world definitions and benchmarks for 'quality' and 'value', it is time for a more positive tone. In this section of the story, we shall discover opportunities for improving the hardware of stand-alone PV systems. We will also find out the essential role that our Visual Instruction System has to play in this. The task ahead is to optimise the hardware as far as possible to the needs of our customers. Furthermore, we must share knowledge to support and strengthen the value proposition that this hardware seeks to provide. First, let us recap:

- 'Value' can be thought of as the primary economic justification for a given product – the most self-evident, compelling proposition for choosing its benefits over others;
- 'Quality' can be described as the degree to which a product is fit for its purpose;
- Dependable, cost-optimised, PV-enabled energy services require an understanding of what PV can do – a fundamental shift in the way it is purchased and a change in the perception of 'solar';
- Stand-alone PV is the highest quality and most cost-effective energy services solution for energy-poor regions, when designed correctly and when used appropriately;
- Autonomous PV is dependent on basic user understanding for operation. More importantly it requires awareness of value, both for purchasing in the first instance, and for optimal operation thereafter;
- PV has specific areas of poor reputation to overcome;
- The obsession with base cost over quality is at the heart of the stalemate in which PV has found itself.

There will always be cheaper products, with ever more implausible claims to their performance. We are forced to accept that PV will only truly progress when everyone in the market understands what we mean by 'quality' and 'value'. I believe that it is possible to foster such an understanding. Furthermore, I shall come to describe how it is even possible to reduce costs in pursuit of value and quality.

Hardware design; Knowledge sharing; Procurement; and Brand. These are the four areas I consider essential to improving the quality and value of stand-alone PV in the developing world; let us explore them next.

CHAPTER 10

Optimising stand-alone PV hardware design

Abstract

Autonomous PV systems in the developing world are typically 10 to 20 times more expensive than they need to be, simply because of how they are specified and procured. In addition, these more expensive systems are generally less reliable and have shorter operational lifetimes.

We are capable of developing more reliable, longer lasting systems which are fit for purpose and cost far less. We can achieve this using a 'holistic' approach to procurement and design.

We define and explore the four principal hardware design considerations to these solutions:

1. *Energy-efficient equipment – the first consideration, and the primary requirement;*
2. *Integrated design – where all components are harmonised with energy efficiency;*
3. *Involving the user – encouraging efficient operation using displays and other notification equipment;*
4. *Data collection – for ongoing design improvements, warranties and customer satisfaction.*

We introduce the concepts of energy ownership through user interfaces, and of guaranteed energy through data collection and pre-failure notifications.

Whenever the subject of solar technology comes up, you can guarantee that the first question will concern its efficiency; closely followed by the battery technology and other specific components. But the most important thing about technical components is how they work with each other and, in particular, how they serve the user.

All too often, stand-alone systems are promoted on the basis of the most efficient PV, highest energy density battery or brightest LED; but these capabilities are irrelevant unless they are used appropriately. The cost of any stand-alone PV system is directly proportional to how much sunlight reaches the PV and the amount of energy required from the battery (however it might be consumed). Both of these factors are dependent on the user. In addition, the performance claims for these products make a number of assumptions, such as correct installation, maintenance and optimal usage.

Two aspects are critical to a PV solution: the system design; and the techniques employed to marry this to the requirements of the application. Both

http://dx.doi.org/10.3362/9781788530705.010

involve making the system as cost-effective and robust as it can be for the energy service it provides, while at the same time engaging the user in the most intuitive way possible.

Holistic solutions

From a technical point of view, it is nearly always better to design and supply a complete solution for a given requirement, rather than for just one section of it. Frequently, PV-controller-battery components are procured as a power supply; intended for operation with equipment selected by others. This often gives rise to several intractable financial and operational problems, the most damaging of which are the high ongoing costs to the user. Historically, solar home systems were culprits of this; mini-grids still are.

The common assumption here is that *electricity* alone is the primary requirement. But the overall electricity requirement is defined by more than the end user equipment's power consumption – it includes the extra energy consumption of the *PV system itself*. Any inefficiency in the system results in wasted energy which is of no benefit to the user – yet it costs money in the form of PV and battery capacity. Autonomous PV systems in the developing world are typically *10 to 20* times more expensive than they need to be, simply because of how they are specified and procured. In addition, these more expensive systems are generally less reliable and have shorter operational lifetimes. This need not be the case. We are capable of developing more reliable, longer lasting systems which are fit for purpose and cost far less. We can achieve this using a 'holistic' approach to procurement and design.

Holistic solutions mean supplying the customer with an entire functional system, from the user equipment to the PV panel and everything in between. There are four principal design considerations to these solutions.

1. **Energy-efficient equipment** – the first consideration, and the primary requirement.
2. **Integrated design** – where all components are harmonised with energy efficiency.
3. **Involving the user** – encouraging efficient operation using displays and other notification equipment.
4. **Data collection** – for ongoing design improvements, warranties and customer satisfaction.

The value of energy-efficient equipment

Take a typical request, such as 'I need a solar power solution for a 1 kW water pump'. On consideration, what they fundamentally want is to move a volume of water from one location to another, in a given time and under certain conditions. Such a request would likely be best fulfilled using a highly efficient and reliable DC water pump designed to work with batteries. (However, as with any project, an analysis of the basic requirements, avoiding technical or other assumptions in the first instance, is crucial to understanding how to achieve a suitable solution.)

Another common enquiry, or written procurement specification, is for a solar energy system to power computers in a school. The example below is a simplified version of a procurement document from the Rwanda Ministry of Education in 2008. (*The total solution costs and the operation without maintenance figures are estimates, but the relative improvements are appropriate.*)

How working computers are typically sourced:

- 20 lowest cost PCs are purchased for, say, US$2,000
- The project operators then look for a power supply solution
- They choose solar for reliability and low maintenance
- They need to use inverters and AC equipment to interface with the computers
- The PV system to power the AC computers will cost ~US$40,000
  - Total solution cost = $42,000
  - Operation without maintenance ~6 months

How PV-enabled computers *should* be sourced:

- 20 energy-efficient PCs are purchased for, say, US$6,000
- The project is designed specifically for solar
- The whole system is DC and solid state – including the PCs (no fans)
- The PV system to power the PCs will cost ~US$8,000
  - Total solution cost = $14,000
  - Operation without maintenance ~5 years

These examples are representative of most applications, from lighting to refrigeration, from telecoms repeater stations to offshore oil platforms. Energy efficiency in the end user equipment is the single biggest factor in optimising cost against the requirements of the application. The resulting solutions are also usually more reliable and last longer: energy efficiency is essential for both quality and value.

Energy-efficient versions of many common appliances are not typically available locally in poorer regions (though this is thankfully changing). However, even with increased equipment cost and international transportation, it is often ultimately much more economically sound to employ this equipment than the cheaper, locally sourced devices. The reality in developing nations is that cheap electrical equipment is often very inefficient and requires far larger PV energy supplies.

The advantages of energy efficiency apply to most applications. Unfortunately, they are not widely practised in the regions for which they hold the highest value.

Energy efficiency in the end user equipment is the single biggest factor in optimising cost against the requirements of the application

Integrated design

So, it is important to reduce energy consumption by the user equipment. But this efficiency must extend to the components within the PV system too. We know that the electrical load requirement directly defines a PV system's cost; but it also influences many other factors such as physical size, reliability and operational life. Here we can bring significant benefits, by removing electrical conversions and other inefficiencies, such as energy loss within cables.

At this point, it is worth mentioning the influence on the PV sector from AC grids, and our continuing, conditioned attachment to them. Off-grid PV dominated the early years of the industry. The real value of these systems has not changed since that time but the attention has shifted drastically away from this sector. This is due in part to an increase in subsidies for grid connection, which has assumed precedence to the point that even PV engineers often have little or no design knowledge of off-grid systems.

A consequence of this is the widespread tendency to include inverters in stand-alone designs. For many applications they are unnecessary, and can dramatically reduce reliability while increasing costs. Consultancy firms without sufficient autonomous PV system design experience often carry this error forward, producing 'standard configurations' for procurement specifications. In the water pump example above, not only will the 1 kW water pump need more energy to move the water compared with an efficient DC pump, but the standard pump requires AC electricity, meaning that an inverter must be used to convert the DC energy from the battery. As a given, any conversion stage introduces more inefficiency into the system; and this will demand more PV modules, a bigger battery and the risk of lower reliability.

There are additional factors to consider. Our AC water pump may have a start-up current significantly higher than its running requirement – a common issue for motors and other electrical equipment. This means that the inverter, certain cables, connectors and switches will have to be sized for the maximum, start-up current, even though they operate most of the time at a much lower level. For the inverter, this means a further loss of efficiency.

Let us ponder some figures for a minute. Let us say that, on starting, it requires 10 amps for 30 seconds; it then runs steadily at 3 amps. Inverters can be over 90% efficient, but usually only at the upper end of their operating range. In our case, the inverter will supply 10 amps at good efficiency, but when the current requirement falls to 3 amps, the efficiency could fall to nearer 50%. From the battery's point of view, this means that approximately 11 amps is drawn for the first 30 seconds, followed by a steady 4.5 amps – 50% more than it needs to be. The battery and PV are also therefore 50% bigger – and 50% more expensive than they need to be.

For the above reasons, using an inverter can account for well over half of the cost of a PV system. The inverter also represents the least *reliable* component, typically with an operational life of 5 to 7 years in harsh environments (compared with 5–10 years for the battery and 20 or more years for the PV and the rest of the system). Inverters often contain moving parts such as fans, as well as capacitors (fast temporary storage) which are generally the weaker points in the design.

The following analysis highlights both the energy and the service consequences of including an inverter in an off-grid PV solution.

Typical AC solution with inverter and inefficient user equipment

1. PV-generated electricity is fed into the charge controller, over 99% of which reaches the battery.
2. After battery charge and discharge losses, 90% of PV electricity can subsequently be drawn from the battery.
3. With an inverter of, say, 70% efficiency, 63% of PV electricity is output to the AC equipment.
4. Assuming that the user equipment (be it a light, a water pump, refrigerator or other) is 70% efficient, **only 44% of the PV electricity is actually used to *perform a service*** – water moved, goods kept cool, light provided and so on.

DC solution with energy-efficient user equipment

1. PV-generated electricity is fed into the charge controller; from there (as above), over 99% of that electricity reaches the battery.
2. 90% of PV electricity can subsequently be drawn from the battery (again, due to battery charge and discharge losses).
3. DC equipment is used, so there is no efficiency loss between the battery and the user appliances.
4. Assuming the user equipment is 90% efficient, **80% of the PV electricity is used to perform a service** – water moved, goods kept cool, light provided.

For the AC solution, both the inverter and the equipment are assumed to have average operating efficiencies of 70%. This may be low compared with some quality equipment, but a figure far lower should be used for the low cost, inefficient inverters and appliances commonly found in energy-poor regions.

Using an AC inverter can account for well over half of the cost of a PV system

Although it is quite common to have efficient DC equipment (because they are designed to run from batteries), it could be argued that the 90% assumed (which makes the numbers neat) makes an unfair comparison with the AC system. However, the proportions are appropriate: removing inverters from system designs is likely to half the electrical size (and therefore the cost) for the majority of 'typical' solutions.

Involving the user

Earlier, in Part Two, I outlined the general principles of involving the user. I want to recall them here with direct reference to autonomous PV applications. Consider for the moment the inconvenience of not having a charge indicator on your smartphone. The user is typically the most important aspect of a PV system, and an agent in its efficiency. Despite this, the technology is rarely utilised to nurture the optimal behaviour from the operator.

Anyone able to influence the operation of the system should be aware of how it operates. One would think this would be a standard consideration (for safety and security reasons, for a start), but amazingly, it is not. Even for fixed-location solutions requiring no actual user intervention, a lack of awareness by those around can affect performance. A real example of this was a large, professionally designed and installed system in Rwanda that began to fail only weeks after commissioning. It turned out that a local woman was washing her clothes and laying them on the array of warm, flat and clean PV modules every morning.

This is particularly important for products which require the user to position and connect the PV panel, such as solar lights and home systems. In another example, a rural African woman returned her solar lamp to a dealer complaining it only worked for three days. The dealer confirmed that the woman had taken the PV out each day and placed it in the sun, as required. After some discussion, it transpired that she had not plugged the PV module into the lamp unit, where the battery was housed. The lamp had provided enough light for two nights without charge, and then failed on the third.

This is not uncommon. Customers of low-cost solar products might never have seen a PV panel, or otherwise understand little of how solar lighting products work. It helps to remind ourselves that PV systems have no moving parts, makes no noise and produce no emissions, so there is nothing to show that they are functioning at all, let alone effectively.

User interfaces

Ironically, given their properties when in proximity, we can draw an analogy between the electrical energy in a battery and a bottle of water. Both resources are fundamental to our modern-day survival. Both are globally ubiquitous, yet locally they are at best a premium, and at worst they are scarce, and require careful husbandry. Crucially, the owner can choose how and when to use the water in the bottle. This is what we require for energy solutions; for its owner to take true ownership, every day. But we have an obvious, innate understanding of how and why we consume water. Conversely, an abstracted, derived form of energy, captured in a box, is not self-evident in what it offers or how best to use it. A man-made process for harnessing nature needs a man-made solution, a helping hand to explain how it works. User interfaces make this a reality.

To recap, a user interface is the input-output mechanism which allows a device to be used by its operator. It is both a control surface and a communicator of feedback. Whether it is a car dashboard, a computer keyboard or a water tap, it must reveal enough of its workings in order to elicit the appropriate responses from its user. For our solar product (which operates automatically), the most basic user interface is a display that conveys information – a single LED showing that *something electrical is happening*. A product may have several of these; others employ LCD or even colour screens. Some LEDs flash when the PV is generating energy, or change colour to show charging or discharging; others indicate charge using battery symbols. They can communicate a range of operational parameters, such as battery voltage, temperatures or efficiency. But as a minimum, these displays should *show the amount of energy remaining in the battery at any given time*. In our analogy, it is like seeing how much water is left in your bottle.

What is most important is that the display 'engages', or places the user at the heart of the system's operation. The universal principles associated with our Visual Instruction System are paramount here. When a display conveys the status of the PV system in a simple and clear manner, very little training is required for the user to learn how the PV system operates in real life. And when this happens, they instinctively take more care of the system, generally making it work more reliably and for longer. *Users increase the quality and value of what they are using.*

Basic PV system operation – an example

Let us revisit the example of a modest PV system powering 20 computers in a school classroom. The system battery is located in a separate room and the PV modules are outside, making no noise and showing no signs of activity. Students use the computers by day, while community members log on for a few hours in the evenings – when lighting is also drawing energy from the PV system. Located prominently on the wall in the computer room is a 20 cm high user display, comprising 20 rows of LEDs in green, amber and red. One row is lit at a time to conserve energy (the display's total energy consumption is negligible).

Throughout the day, operation proceeds as follows. The battery's state of charge (SOC) is lowest just before dawn; though under normal operating conditions the battery should have over 40% capacity available (required for good battery health and life). Our display, therefore, starts the day illuminating the bottom of the amber LED rows. After sunrise, the PV modules start to generate electricity, and the display begins to illuminate the next row up each 15 or so minutes, to reflect the increase in battery charge. As the students arrive and switch the computers on, the display remains unchanged. Here, the computers are drawing power, but the battery is being charged. Throughout the morning, we see a net increase in charge, until after midday, when the display moves into the green rows of LEDs.

The students finish school mid-afternoon and – knowing why it is important – switch off their computers. The display continues to climb the green rows; despite the now-reducing sun energy, there is a reduced energy demand by the computers. The battery reaches its highest charge level at the end of the afternoon. Charging stops at dusk; and the lights are switched on in the classroom. Being energy-efficient, they have a minimal effect on the battery charge, but as computers are turned on in the evening, the display reflects a steady reduction in available energy. By keeping an eye on the display, no terminal suddenly runs out of power, and some can be prioritised over others if the display falls into the amber rows.

No understanding of watts, amp-hours or voltages is required in this scenario. After becoming familiar with daily system operation and the behaviour of the display, the user will be able to estimate the remaining battery energy in terms of time, noting how many computers are being used, and the sunlight conditions for that day. Such remote displays have many applications: solar home systems; medical facilities, particularly critical devices such as vaccine fridges; businesses (computers, internet access, lights, fans, radios and more); and of course classrooms of computers, where the display can also be an education tool.

People in emerging economies exhibit naturally energy-efficient behaviour, once they understand the practical dynamics of the equipment, and their own stake in using it. Of course, the profligacy of the grid-conditioned rich nations stands in stark contrast, but here too, we notice the likes of wearable fitness technology doing *exactly* the same thing, providing a simple visual incentive towards vigilance in matters of health and exercise. User interfaces are our means of tapping into the naturally prudent mindset of the end user.

People in emerging economies exhibit naturally
energy-efficient behaviour, once they understand
the practical dynamics of the equipment

One of the key value propositions for these systems, then, is for people to have energy ownership. Our example of a display offers freedom to choose which services are most important. But they can also be set to represent important energy storage thresholds, depending on the product. Even simple three-LED displays of green-amber-red can be used to add a further level of awareness for the user; as well as the basic *amount* of energy available, it can show changes in the *rate of use* over time, thereby feeding back a measure of *efficiency* to the user. What is now being communicated is a function of its operation; in other words, we are communicating *value*.

Example 1: A solar home system that is powering several appliances including a fridge. Condition: There must be a minimum amount of energy at the end of the day so the battery can still support the fridge until dawn/charging. There also needs to be spare energy for days of bad weather and in winter/rainy season.

- Green = within daily energy allowance
- Amber = using more that day than allocated
- Red = system is low on energy and will cut out soon

Note that this scale communicates two different things; the amount of energy available and the rate of use.

Example 2: A solar home system that is powering several appliances including a TV. Condition: If the TV is viewed all day, there may not be enough power for lights at night.

- Green = lots of energy in the battery
- Amber = there is only 4 hours' worth of light energy remaining
- Red = there is 1 hour of light remaining

This scale, on the other hand, shows only the amount of energy available.

Displays can be very simple, and are very low in cost. Yet so often they are omitted from products, or compromised to save a dollar or less on the sale price. Some suppliers fail to include them when they have negligible cost compared with the overall system. Even for small solar home systems where the interface could count for 10% of the overall solution, it is still worth the money, because of the value it presents to the user.

The enormous value of recorded data

Earlier we saw how holistic approaches to design address cost and efficiency. An altogether more difficult thing to control is reliability. We aspire to reliability as a core value for stand-alone PV, over other electrical generation technologies. Without it, generic PV will never be taken seriously as a foundation for the millions who need dependable energy services.

Poor performance usually takes the form of *lower available energy* to the user; fewer hours of lighting, TV, radio and so on. It also typically results in a much shorter battery life – sometimes months instead of years. The reasons for this range from low-quality components, under-sized design specifications on grounds of cost, energy demand in excess of the system's design; or local conditions, such as weather, shading or dirt. In most cases, it is very difficult to prove *why* a system has underperformed. However, if we can record key parameters of the system, then we are able to deduce the causes of a given failure; in fact, we can anticipate failures before they happen.

Data collection, also known as data logging, involves measuring certain system parameters – such as PV output, battery output and ambient temperature – and saving the data into memory for monitoring and troubleshooting. This is a mature discipline, employed in thousands of applications around the world. How does it work?

Any PV system charge controller product which uses a microprocessor can perform data logging, and many of the leading charge controllers on the market for larger PV systems now integrate this feature. (External or stand-alone data loggers can be used, but they either draw energy from the PV system or have their own battery. We would prefer to avoid these scenarios.)

Firstly, the microprocessor saves system data into memory in real time. It then time-stamps and encrypts the information to ensure that it can be extracted, well-formed, without corruption. The information is retrieved by connecting directly to the controller/logger using a laptop computer or USB storage device, or otherwise remotely. The memory capacity is usually large enough to store several years' worth of data. As a minimum, the memory should hold all performance data within the warranty period of the equipment; but should be expandable to allow a longer data record, or a higher resolution of capture. Ideally, the storage medium should be non-volatile, which means that it requires no additional energy to store the data, and will not be affected by system failures. Just as with common USB and digital storage memory, data can be stored indefinitely. Once extracted and decrypted (which need not be a highly skilled task), it can be imported into a standard spreadsheet application to allow easy graphical representation of the system's performance over time.

The value of data collection for stand-alone PV

For a stand-alone PV product or system, the battery nearly always defines daily energy availability and operational life. It is often only through monitoring and data analysis that the specific causes of a battery failure can be deduced.

This fact lies at the heart of all the poor perception of PV. You can install a system and demonstrate that it operates perfectly well when the battery is new; but too often the cheaper PV or battery will begin to reveal the limitations in their design or deployment in the longer term. We have seen how the generic scapegoat 'solar' is commonly singled out for blame here, instead of the unrealistic claims made by individual suppliers or manufacturers.

Data logging and presentation of data is one way to overcome lowest-cost purchasing. For lead-acid batteries (used for the majority of larger systems) the voltage is a close representation of its state of charge – the amount of energy it is holding. (Different, but equivalent data is readily available for lithium and other batteries.) By analysing battery voltages at dusk and dawn, we can assess the amount of electricity fed to the battery for a particular day. Compared against the weather conditions for that day, we can assess how well the PV array is charging the battery. If charging is poor on a sunny day, then either there is something wrong with the PV (which can be verified visually), or the battery is not accepting charge very efficiently, an indicator that it is old or damaged.

Extending this type of analysis over longer time periods, we can gain an understanding of the health of the system as a whole. If, for example, the battery voltage falls quickly between dusk and dawn, over a period of months, it shows that the battery is approaching the end of its life. It can be replaced before its capacity becomes too weak to operate normally.

From an operational perspective, data collection can help to ascertain the following:

- The condition of a battery when it was installed (damage from transportation and installation)

- The total amount of power drawn each day (whether the system is being used correctly);
- The rate at which power is drawn (which affects battery capacity and life);
- The temperature in which the battery operates (this can drastically affect operational life);
- How well charged the battery is at the end of each day (PV array performance);
- How deeply discharged the battery is at the beginning of each day (controller functionality);
- The rate of charging (which is constantly changing during the day and affects charge efficiency).

The broader value of data collection

The role of data collection goes beyond the merely technical. It has a broader promotional and legal remit, which support due diligence, knowledge sharing and, ultimately, reputation. In this capacity, it can be used for the following:

- Proof of performance for future contracts and promotional purposes
- Optimum sizing of systems based on real, measurable data
- To aid product fault finding and future design
- Monitoring of load equipment consumption to detect possible faults (telecoms, public facilities, solar home systems, lighting, PCs, etc.)
- Training of users and interested parties in PV system operation
- Warranty claims by the customer or defence by the supplier
- Carbon credits and carbon offset legal verification

I have seen dozens of procurement specifications for solar powered equipment which include requirements for how funds are spent, yet hardly any for assessment on the performance of the PV equipment after it is installed. This effectively leaves the shed door open at night. A low-cost bidder not only wins the initial supply contract; they are subsequently *paid* to repair the equipment when it fails prematurely. The customer can be told that the product's lack of performance was down to user error, with no legal retort. This is how funds for educational, medical and 'poverty alleviation' energy projects are haemorrhaged.

Data collection offers value truly in excess of its cost. In one of the charge controllers I designed for use in the UK, the data logging hardware content was approximately £2, 3% of the total cost. There were obviously development costs for the software and hardware; but, even when amortising these into the product production cost, the value remains incredibly high. Its greatest value is that it can provide confidence to those needing dependable energy. With appropriate local supplier support, it can help enable guaranteed energy for a customer. We are seeing more and more system controllers with data logging capacity (at little or no premium), so it is now the role of the local suppliers to utilise the facilities and massively strengthen the value proposition of the systems they provide.

Hardware design: a summary

1. Energy-efficient equipment may be more expensive to purchase, but it is likely to save significantly more cost for the PV power supply.
2. Integrated design – removing electrical conversions and short life components – reduces cost further and increases reliability.
3. A complete DC solution is safer and more reliable than one with an AC component.
4. Involving the user improves system reliability; in many cases, it massively increases its value.
5. User interfaces are low-cost, high-value assets which optimise value for the user. They also enable a high degree of versatility in how the energy is used.
6. Data collection provides proof of what PV can do for those needing a foundation for their energy infrastructure.
7. User interfaces enable energy ownership. Data collection enables guaranteed energy services.

CHAPTER 11

Sharing essential knowledge with everyone

Education is the most powerful weapon which you can use to change the world.

— Nelson Mandela

Remember, you can always find East by staring directly at the sun.

— Bart Simpson

Abstract

Autonomous PV solutions, if they are to be effective, require stakeholders, understanding. These individuals, volunteers, commercial buyers and government procurement officers truly need to appreciate value *– that these systems are only reliable when designed and supported correctly, and that the cheapest solution will not necessarily deliver value, or fitness-for-purpose.*

Promoting quality and value prompts us to address PV's tainted reputation; in particular, the gap between product claims and real-life operation; how to get the best from PV; and how to choose the highest value.

The central role of our Visual Instruction System is to teach quality and value to tens of millions of individuals with diverse circumstances.

My message is simple: visual media – pictures, animations, storyboards – stands as the most important investment a supplier of stand-alone PV products can make. It is at least as valuable as the product itself.

My early years of PV engineering were with a small UK company which supplied bespoke systems for industrial applications, such as telecommunications towers and oil and gas infrastructure. We delivered systems with 30-year lifespans, often with the equipment operating in extreme climates, and in some cases, explosive environments. Our customers considered PV systems expensive; but they also knew it was the most reliable form of electricity supply.

At the same time (circa 2002) I followed with interest an education programme in Malawi, central Africa. Here, stand-alone PV solutions were repeatedly dismissed as too complicated, expensive and unreliable. I visited Africa many times in the years that followed, trying to challenge this perception: conducting training in how to operate and maintain simple PV systems; writing proposals to procurement organisations; and meeting with local solar companies to show what was possible. Slowly, I accepted that this dismissal of solar, based on its reputation, was in fact justified.

http://dx.doi.org/10.3362/9781788530705.011

Previously, I had worked with people who had already identified PV as their solution, or those who had already been provided with solar. Teaching how solar works – how to use it, what it can do – was bottom of my list of priorities. Now, with the frustrations of the Malawi project still gnawing, my priority reversed. The road to understanding the root causes of this bad reputation has been a long and complicated one. But one fact has become clear: the most fundamental hindrance to the uptake of stand-alone PV is a lack of understanding of what it can do.

People we aim to reach

If we want people to appreciate the value of autonomous PV systems and utilise them, it is useful to consider who these people are; the primary groups of people we need to inform and educate; in short, our stakeholders. I have deliberately used basic terminology to avoid being over-analytical about these groups, but I have ordered the list with care.

- People using PV products and systems
- People supplying PV
- People making PV components and products
- People specifying PV
- People financing PV projects and initiatives
- People who could use or need PV

So far, we know that stand-alone PV comes into its own as a means of providing a complete energy service, rather than a mere power supply. Furthermore, you can only hope for a solution of any real quality if you design it against a set of known requirements. (We have used the word *holistic* to describe such solutions.) These requirements can only be defined by *the people who actually need the energy services*. This poses a problem if, as in many situations, PV is not even considered as a serious possibility. In places without reliable mains electricity, it is vital that our stakeholders understand that stand-alone PV is the most dependable energy service solution. What is more, these individuals, volunteers, commercial buyers and government procurement officers truly need to appreciate *value* – that these systems are only reliable when designed and supported correctly, and that the cheapest solution will not necessarily deliver value, or fitness-for-purpose. At present, if an individual wants mobile phone charging or personal light, or if a nation needs a Wi-Fi or TV network, stand-alone PV is typically only a secondary 'alternative' consideration. My aim is to raise its status to the *first* question for anyone in the developing world. Put another way, why would you not use PV?

The category at the bottom of our list – people who could use PV but have no access to reliable mains electricity – represents well over a billion people. You will appreciate that this is too large a sector for us simply to launch a marketing campaign or conduct a training programme. Promoting quality and value prompts us to address PV's tainted reputation; in particular, the gap between product claims and real-life operation; how to get the best from PV; and how to choose the highest value.

This is not just a case of targeting specific groups or professions. This is knowledge which has to be propagated to tens of millions of individuals who share neither language, nor culture, nor literacy levels, nor technological awareness. For me, realising the scale of this undertaking was another one of those landmark moments where the myriad challenges were thrown into sharp and instant focus. For a start, the need to circumvent written language, while also raising awareness, improving reputation, fulfilling potential and promoting value.

It is vital that our stakeholders understand that stand-alone
PV is the most dependable energy service solution

Autonomous PV solutions, if they are to be effective, require user understanding. This is in short supply, which is why stand-alone PV has had limited impact on poverty alleviation over the 30+ years it has been deployed. It will continue to hinder generic PV in the future unless we wean ourselves off sales brochures, performance claims, user instructions and value arguments that largely comprise text.

Word of mouth and English as a foreign language

In poor countries, word of mouth is by far the most common form of information transfer. Although mobile phones would, on the surface, appear to be changing this, you are likely to find a village where a single product is discussed among hundreds of people. A person visiting an urban area will return with stories of what they have seen. Sensitivity and judgement are needed here. It is common practice that many goods of value to rural life are provided free of charge to influential individuals, so that news will spread quickly. Religious elders, political figures and community leaders are all targeted for this role. Unfortunately, this is often a bad thing for PV because its value is so dependent on correct operation. A well-intentioned but incorrect demonstration of a PV product serves to spread misinformation about its capabilities or its operation. The dynamics of groups doesn't help. Even a member of the group who recognises a public mistake might not feel permitted to speak out. People are people: pride can interfere with learning, and misinformation can go uncorrected in groups.

If only there were instructions

Until very recently all PV product user instructions and packaging were written in a single language, usually English. I have raised this issue with a range of people. Rural communities in the hills of Rwanda, Indian distributors in Kenya and manufacturers in China have all confirmed problems from the use of unfamiliar text on solar products and packaging. While most end users of

PV in the developing world do not read English, many may not read at all. In places where literacy itself presents a basic challenge, illiterate individuals can see written text as intimidating and as a barrier, and may consider themselves belittled. It is worth considering the unease with which sectors of industrialised societies adapt to new technologies in this respect. If the product does not look easy to use, senior figures within a local social hierarchy may not even try to use it for fear of losing face. Worse still, they may discourage lower members from attempting to use it. The success of the product therefore relies on the user instructions and the seller. If there are effectively no instructions, this leaves the seller as the sole carrier of the product's fortunes. This is not altogether desirable.

The local level

Users of solar products typically buy from local retail outlets or local distributors. (There are brand-specific supply chains, but as we shall see later, these are relatively short-lived and are not commercially justified at the required global scale.) Naturally, these sellers want to make money: they are not concerned with teaching customers how to use a solar product (even if they actually knew how). Additionally, sellers raise expectations by over-egging the product to make the sale. This, however, is the way of the world; at this stage of the game we are not in the business of changing human nature, or denying a livelihood.

For most distributors and retailers, 'solar' is just another product. Obvious, perhaps, but this presents us with a problem. All consumer products require a degree of user understanding, no matter how well designed they are. The difference is this: the best designed solar light with all the latest technology *will not function at all* if the user doesn't know to put the PV in the sun every day and plug it into the battery unit.

Distributors do not sit down and study the details of the many products they sell. They might already have electricity in their shop or premises and have no personal need for a solar light. The problems of distribution and education have been recognised for many years, but they are difficult to address. The recent period 2012–2014 saw several initiatives to strengthen solar-aware distributor networks by individual commercial companies and organisations, such as the UN.

But I consider this largely wasted money and effort. With respect to a self-sustained local PV industry, and mindful of a future for the countries that need it, I believe that we should acknowledge these supply networks; accept that they are truly dynamic – in terms of people, focus, scale, speed and diversity. Instead of trying to tailor the supply network to our products, wouldn't it be a better approach to format our products to the existing supply networks? This entails utilising existing distributors of every size from those handling the thousands to the person selling a single unit per week. This serves to remind us of our target market, our stakeholders and, ultimately, our requirements. We will return to the challenges of local suppliers later; but for now, users and supply chains (local or international) all need information which is quick to understand, clear and versatile, with no written words.

The value of visual media

This is why I am convinced that packaging, sales literature and user instructions are not peripheral add-ons to the consumer product. *Their role is to teach quality and value.* A Visual Instruction System – an intuitive, symbolic communication protocol – is at least as important as the PV product or solution itself; it should be seen as synonymous with the PV solution, and merit equal status in matters of research, development and budgeting. This is a necessity for the widespread uptake of solar PV – and, I contest, the wider effort to end energy poverty.

My message is simple: visual media – pictures, animations, storyboards – stands as the most important investment a supplier of stand-alone PV products can make. Here is the logic.

Our target customers in energy-poor regions (2 billion people) cannot understand written sales and marketing material. These people speak thousands of languages, and rarely English. Illiteracy is widespread.
Conclusion: words are meaningless (and possibly detrimental) to most potential customers.

Stand-alone products, however good, require user understanding for correct operation and maintenance. (Badly situated PV, for example, seriously compromises the performance of the product.) Poor performance (even where due to user error) reflects badly on the supplier.
Conclusion: suppliers need to explain to customers how to use products.

There will always be cheaper products with inflated performance claims but which fail to work. Suppliers wanting sustained commercial success need people to understand the benefits of a product which is fit for purpose.
Conclusion: suppliers need to explain value.

Good products have the most value among rural populations. But these potential customers are widely dispersed geographically. It is therefore expensive and time consuming to visit even the large rural communities. Suppliers cannot afford to even train sales personnel to visit customers, let alone pay people and their costs.
Conclusion: it is not possible to visit all potential customers.
Summary: Most potential customers cannot read, but suppliers need to explain value without visiting them.

Visual media can be distributed to all potential customers at very little cost. The leap of faith for investors, designers, manufacturers and suppliers is that this is a production cost; but the financial burden of visual media is all in the initial production, after which it yields value for everyone.

Our stakeholders, meanwhile, can identify better products. Opting for quality, they will feel more confident when challenging the lowest-cost orthodoxy. Understanding how to look after the product yields further value. Designers and manufacturers can place more emphasis on visual media knowing that it will increase sales, reduce product returns and warranty claims, and support ongoing business.

The financial burden of visual media is all in the initial
production, after which it yields value for everyone

Training and investment

All reliable, fit-for-purpose PV solutions require technical input. No matter how good a PV product or system design might be, its performance is ultimately dependent on how it is used, its environment and its application. For the most popular PV products, such as lights and phone chargers, it is not easy to provide the kind of user support that teaches 'value' as we know it; it is trickier still for larger, multi-purpose solutions.

Such solutions demand structured training, and a workforce dedicated to the local autonomous energy services industry. At the very least, a user must appreciate the principle that the battery only has a limited amount of energy; and that this is dependent on the devices connected to it, as well as the sunlight's intensity. Where a PV system offers several potential applications, only the user knows how to gain the highest value. Here, then, if we can convey the advantages of using the products efficiently, they can enjoy true 'ownership' of their energy system, deciding for themselves which service is most important: TV or fridge, radio or fan, phone charging or light.

For anybody reading this as an instruction for sucking eggs, or alternatively if you think it too expensive or too demanding logistically, step back a second and consider. Training, as we envisage it here, is anything but a throwaway expense. Underpinning all of this is a business model – just built on different foundations of people, quality and value. Knowledge sharing is a true investment, whose return is a unique opening into a vast and otherwise impenetrable market, available to all. If people of all geographical locations can earn money and hold value to suppliers, those foundations and lines of communication are already in place for training and teaching to be made available.

The requirements for training materials can be summarised as the following value areas:

- **Technical**: installation and maintenance; commissioning and fault-finding; and PV system design;
- **Commercial**: identifying opportunities; value propositions; funding; and sales.

And so, word gets around of services tailored to the needs of the user or application and used anywhere. As more people get to know of stand-alone PV's capabilities, we will see a greater uptake.

Public confidence

Traditionally, designers and suppliers have tried to figure out the requirements of a given product by a mixture of formal and informal means. This requires insight into the breadth of potential applications in a given area, and the needs of users. But as we have noted, PV is dependent on something more fundamental: the innate endorsement by the public that it is a credible option, by virtue of its scalability, zero running cost, low maintenance requirement and silent operation. Our stakeholders should come to know that in basic terms, stand-alone PV can be as dependable in the developing world as mains electricity is in industrialised countries.

We seldom ask how the mains network in Europe is kept in operation 24 hours per day; we just know and are satisfied that it is. The same goes for electrical appliances: we know a television or media player will operate for several years, but we don't ask what it is about the design that makes this the case. Similarly, in building the brand and general reputation of PV, we must aim for a brand reputation that is impregnable; that PV is simply a dependable energy services solution. Its ultimate brand value is that it can be guaranteed and almost maintenance free – as long as you buy a quality product from someone who knows what they are doing, and understand how to operate it.

In order to achieve this, we have to bring knowledge of the potential solutions to the people who will sell, use, install and maintain the equipment: it is an intrinsic part of our product and brand development.

Our stakeholders should come to know that in basic terms,
stand-alone PV can be as dependable in the developing
world as mains electricity is in industrialised countries

- We need to convey the dependability of PV to more than a billion people who can benefit from dependable energy services;
- We have to communicate in an engaging and universally accessible manner, without using written language;
- We must employ visual media – drawings, animations, storyboards, photographs and iconography, for users, suppliers, stakeholders and influencers;
- We must incorporate local people, particularly in remote and rural regions, into sales and support networks;
- We must provide training and teaching, for dynamic support and to identify local opportunities;
- We must ensure that PV users appreciate how to draw the greatest value from their solar energy service;
- We must provide public information, for reputation building and autonomous PV value awareness.

CHAPTER 12

The need to reform procurement of stand-alone PV projects

Abstract

For various reasons, formal procurement processes for stand-alone PV solutions in the public domain, far from ensuring value and quality, have pretty much become synonymous with sub-standard reliability and operational life; and rocketing costs.

Approaching procurement specifications in the way I describe in this chapter will result in high quality and high value energy services for the end user. Overall project cost will often be far lower.

Operational costs for the supplier will be lower, the system can be maintained more easily and, most importantly, all stakeholders can have confidence in the energy services being provided. This is where the true value of PV systems really lies.

As far as equipment is concerned, the specifications will optimise the cost of PV solutions. The equipment is safer (electrically, mechanically and chemically speaking) for installers and users alike. The environmental impact, from installation to the end of the system's life, is significantly reduced.

Our future holds the potential for a long-term, sustainable energy infrastructure with a solar foundation. My belief is that such a future, if it is to happen, must be commercially driven, by which I mean that it is profitable for everyone in the supply chain.

It's not as if the money isn't there. It's just that the visible demand is being served by short-term solutions. As you read this sentence, the gargantuan sums being thrown at unreliable mains and disposable batteries could easily support massive programmes of autonomous energy services.

The shortfall, as we have noticed so far, lies elsewhere: in the reputation garnered during the PV industry's growing pains; and in the awareness, initiative and collective will to overcome this. I have described how technology can enhance quality and value, by way of rigorous design and user involvement. Instruction and support using visual media can inspire confidence and understanding of PV's capabilities. Also, it is necessary to tailor PV products and supporting media to existing supply networks, raising awareness and appreciation from the ground up.

http://dx.doi.org/10.3362/9781788530705.012

Talking to the right people

This is all very well. Unfortunately, everyday people and professionals do not manage the large projects that are so critical to modern energy services provision. Sadly, whatever demand exists on the ground seems to take an enormous amount of time to bear fruit in any sort of formalised procurement.

By far the most influential people we need to reach are the financiers of energy services. This is because the funds for large projects and initiatives – lighting or home systems, electricity for schools, healthcare or rural electrification – are usually from industrialised nations or bodies. They nearly always originate outside of the country where the services will be used.

In an attempt to ensure the money is used appropriately, the funding is only allocated against a detailed specification for the equipment. Whether it is an NGO, large commercial supplier or international donor, the technical specification for the actual equipment typically lies deep within the procurement documents.

Procurement specifications

Written procurement specifications have been used for fixed-location stand-alone PV systems since their first application in the 1970s. Generally, these installations enabled services which were critical in some way: communications, safety, monitoring, flow control and so on. They were often expensive, requiring rigorous formal design prior to any physical implementation. The client would issue a technical specification to which all suppliers would have to comply.

In the early days, the solar power specifications were actually developed in conjunction with the PV system manufacturer, because they were the only companies able to design autonomous PV to the reliability levels needed. In time, these specifications became established within the client organisations. For industrial PV systems (be they large, fixed location or small distributed monitoring solutions), these specifications have remained the standard model for procurement. And they work well. For the last 30 years, PV systems have been routinely supplied with lengthy operational warranties and design lives in excess of 20 years. But more recently, procurement, particularly in the public domain (for example, urban solar street lighting and rural electrification), has created a fundamental barrier to the long-term success of PV. For various reasons, formal procurement processes, far from ensuring value and quality, have pretty much become synonymous with sub-standard reliability and operational life; and rocketing costs.

This is a serious and growing problem. Until recently, these detailed procurement specifications were only employed for high-value or high-volume requirements. However, these ever-more complicated documents are being used for smaller and smaller procurement budgets, meaning larger numbers of projects. Overall, they are wasting eye-watering amounts of money.

An example of procurement: Uganda Ministry of Education and Sports

Let us look at a fairly recent example: 310 PV systems procured by the Uganda Ministry of Education and Sports (MoES), for what it called 'Post Primary Education Institutions'.

The 'invitation to bid' documents consisted of 298 pages, the bulk of which was related to commercial terms and conditions, but also included the PV technical specification. This project document pack was typical of these monster specifications. There are, broadly speaking, three layers.

The first layer is usually the biggest, and defines how the tender process must operate (such as instructions to bidders; how the bids will be evaluated; eligible countries; data sheets to be completed by bidders; and so on). This is fairly standard content, almost certainly originated by the World Bank or UN.

The second layer is concerned with the 'employer' or national procurement body – in this case the Uganda Ministry of Education and Sports (MoES). This contains information on timescales, MoES contact details and detailed specific World Bank compliance instructions, in the form of templates to be completed by the bidder.

The third level concerns the actual project – including very detailed information on everything from the recipient schools, specific locations (including climate data) to the full equipment component list and warranty requirements.

The Uganda project equipment list specified PV array size (and rating of each module); battery type and capacity; our dreaded inverters as well as 'balance of system' (BoS) items, such as cables and metalwork. In short, all the design options had been removed. Energy-efficient equipment was not permitted. Neither was any form of integrated design (a way to avoid using those energy-sapping inverters). No user displays were specified; and aside from standard warranty terms, there was no mention of post-commissioning performance – which meant that there was no provision for data logging.

I was a consultant for a Chinese PV manufacturer at the time. I made some rough calculations for a fully compliant technical bid, *versus* what I considered a common-sense solution; that is, delivering the same performance but more reliably, with longer life and with lower maintenance demands. The compliant bid added up to US$13 million; my optimised solution came in at approximately US$3 million.

We did not submit a bid. The documentation stated 'alternative bids are not permitted', in bold, with a follow-up paragraph reinforcing the statement. If we had made a compliant bid, we would be competing mainly on price; so margins would be low, for a project with significant potential for unforeseen costs – particularly in logistics. It was not worth the risk.

The pros and cons of procurement processes

This procurement method is the norm, not the exception. It is not uncommon for tens of millions of dollars to be wasted within these procurement processes. (I am suddenly shocked at how casually I wrote that sentence.) The end result is unreliable and sub-standard *by specification* – which, among other things, reflects badly on PV as a generic technology.

Such failures are down to much more than the excessive expenditure on equipment; often the deciding factor is the time and money spent on bureaucratic assessments, which result in frustratingly obstructive procurement requirements. The ramifications are far-reaching. National and international organisations – such as farmer co-operatives, NGOs, grant bodies – often follow the basic structure of these procurement specifications, maybe believing that this provides a degree of confidence, credibility or justification for funding.

Often the deciding factor is the time and money
spent on bureaucratic assessments, which result
in highly obstructive procurement requirements

Hundreds of solar procurement projects are initiated through these specifications, in the fields of education, lighting, healthcare, communication and safety. Each one is likely to waste money and produce equipment solutions that fail to meet the needs of their intended beneficiaries.

Legacy specifications

You might be thinking that all of this sounds a bit improbable. Can this really be happening – and on such a scale? Looking at the specifications, you may think: 'There *must* be some reason for this?' 'There must be a good reason for spending so much money, for going to such inflexible detail?' But trust your instincts; I don't think there is. Instead, I think that much of this is historical. Between the lines, I often see requirements which *could* have made sense many years ago, but which make absolutely no sense now. A particular example from the past demonstrates this issue well.

In my industrial PV days, our systems for the oil, gas and telecoms sectors were all supplied against comprehensive technical procurement specifications. Among the items that always used to crop up was a requirement for metal battery enclosures. Large lead-acid batteries require careful handling and installation into enclosures to contain dangerous leakage in the event of physical damage, while resisting explosion from within. At the time (around the early 1990s), metal boxes were the only format strong enough to serve this purpose. They were lockable and they were pretty solid. But they were stainless steel (to resist environmental corrosion) and very expensive.

During a bid in 2001–2002, we approached our client with a recommendation for GRP battery boxes (glass reinforced plastic – the material used for products such as surfboards, boats and roofing panels). The material was well proven and the boxes acid-resistant. They would contain a battery explosion by bending and absorbing the force in the manner of Kevlar bullet-proof jackets. Additionally, these non-metal containers presented no short-circuiting risk from the battery cells. They were as safe as stainless steel, but far lighter, easier to transport and much cheaper. Once it was confirmed that the GRP versions were compliant with the needs of the project, the client accepted the choice of material. Surprisingly, they then revealed that they *knew* the technical specifications were out of date, but they were very hard to revise. In-company specialists and external consultants had originally produced the documents under a very arduous process, refining them over the following months before they were set in stone. They could not be changed without being ratified by someone external to the client company, with PV energy expertise.

We, as a specialist PV system supplier, engaged to modify the specification. We presented our recommendations clearly and continued an open dialogue with the client. Our updated version of the client procurement specification was approved a long year and a half later, after the client had sought confirmation from a third-party consultant and ruled out any design conflicts with other aspects of their equipment hardware.

International donor specifications

The World Bank and international donor specifications show similar signs of age, but they do not have the flexibility to allow a dynamic modification process.

I must state some sympathy here. Suppliers are often simply trying to cut corners to save logistical time and make more profits. Sometimes the implications of a design change may escape them. The organisations spending the money risk a great deal by having their specifications modified from the original, approved version. Central to the problem is that most of the procurement personnel are not technical. In addition, the technical teams are often academic rather than industry veterans. This means they are less likely to have an up-to-date working knowledge of the scope of possibilities in terms of design, sizing, component products, selection of materials or application techniques.

In both the industrialised and developing worlds, specifications contain tell-tale assumptions in matters such as the capacity of the PV panels, battery capacity and the use of particular types of inverters. These show an attachment to history that is no longer relevant or appropriate. For example, DC load equipment is now commonly available. In the past, you needed an inverter for most applications. No longer true – and you absolutely don't want them near your PV solution, devouring cost and reliability. This is true for DC communication equipment, low energy lighting and sterilisers for healthcare. These applications, and hundreds more like them, can be made very efficiently and are widely available. However, many did not exist 10 years ago, when the specifications were in development.

Specifications show an attachment to history
that is no longer relevant or appropriate

How do you get to modify these documents? There are no quick or easy answers. It proves difficult to find someone to meet the bureaucratic challenges within these giant organisations, let alone someone fit to manage the actual modifications. Smaller agencies or organisations look to the likes of the World Bank and UN, and risk becoming non-compliant for certain projects if they dare act independently upon their procurement methods.

Nevertheless, act they must. All of these PV procurement documents need to be both modified and modernised to keep up with technological change, if they are to stand a chance of reducing costs and benefitting users. Meanwhile, donors in all their different forms, from well-meaning individuals through to governments and international agencies, are spending excessive amounts of money on unreliable sub-standard technology, diverting investment from the education or the health of the people they are supposed to be helping.

Dynamic modification: readiness for change

Industrial procurement, then, differs from international donor procurement. Not so much in terms of the level of detail; after all, both insist upon diligence and due process; and in both cases we find comprehensive equipment requirements. Rather, it is the *dynamic* nature of the procurement which sets them apart.

Crucially, the industrial documents allow bidders (specialists in PV system design, remember) to propose alternative solutions. They do this by prioritising the overall deliverables above everything else; *the installed solution must perform for X years in Y environment with Z maintenance*, or something along those lines. This established, the supplier is legally bound to the overall requirements: if the system underperforms, they must address the issues.

The international donor specifications (at least, those that I have seen) simply do not allow this. The World Bank alone, prominent in many developing world solar initiatives, has spent over $1 billion since the mid-1970s, and from most perspectives – say, tackling poverty, raising education level or rural electrification – nowhere do the results reflect the level of spending. I am convinced the majority of these procurement processes lead to low quality solutions, and a waste of money.

Even if the giveaway historical fingerprints of these documents were not compelling enough evidence of administrative stasis, the reality is that procurement specifications work in industry, but lack effectiveness in societal applications. It is the dynamic nature of the procurement and a focus on delivered service that nurtures success in industrial projects. The client exploits specialist knowledge, rather than suppressing it through a well-meaning but bungled bureaucracy. This is why I assert at the outset that a *commercial* incentive is at the heart of our mission.

If this raises a few ironical eyebrows, consider this. Yes, we have seen the damage wrought by low-cost, profit-motivated solutions (cheap stuff that doesn't work). But we have also explored how to bring quality and value to the table; a weightier commercial proposition, carrying a degree of responsibility and cost; but which promises rich rewards.

Procurement specifications work in industry,
but lack effectiveness in societal applications

Warranties and realities

While not exclusively so, by and large the big, internationally supported projects for health, education, communication and poverty alleviation – the ones with the greatest potential for good – are procured by national governments. As noted earlier, these programmes are funded from overseas organisations such as the World Bank, foundations or industrialised nations. As well as the time-honoured practice of lowest-cost purchasing, in the interests of transparency there now exists a sizeable body of due diligence, to verify how the money is spent; but there is little post-supply analysis. Even when local officials recognise these issues, it can be difficult to feed back to the donors through official channels.

In principle, governments procure solutions with warranties, but they are often not enforced. I have seen numerous examples in Africa of PV systems supplied to a school with contract specifications of 5 years. These cheaply quoted systems typically fail within 2 years, due to poor design, inadequate PV capacity or battery technology. As an engineer of stand-alone solutions in the public sector, I am constantly disappointed to see specifications that fail to uphold reliability or value, and I am regularly contacted for advice about PV systems that have failed prematurely due to poor design.

It is clear from conversations with local suppliers and engineers that many genuinely believe their systems are designed appropriately. They are simply not aware of the finer details of design – such as temperature effects, battery ageing or voltage reduction in cables – which reduce the life of the overall solution.

But others know the system will under-perform. They also know that if they are not the cheapest bidder, they will not win the contract. And as we have seen, suppliers are often paid to fix the very system they should have designed correctly in the first place. It is rare for a track record to be held against suppliers, and so the cycle repeats.

So, what can be done? A major step in the right direction is for local government to procure the *total* solution, instead of discrete system components, and to focus on the *actual needs of the end user* when issuing procurement specifications.

The needs of the project

In our real-life example, the requirements of the Uganda Ministry of Education and Sports were very simple: they wanted 20 computers to be used 8 hours per day, for 5 years, in specified locations. They wanted the computers operational (not just the power supply) and this should be the responsibility of the supplier.

Let PV system designers utilise the best
technology and approach for the task

So why were they stipulating technical criteria for a specialist energy technology? Clear, precise language regarding the actual service requirement – *reliable computers* – should have been the principal theme of these specifications. Let PV system designers utilise the best technology and approach for the task. If the supplied solutions fail, the supplier fixes them at their own cost. Failures of support can incur exclusion from future procurements (of all kinds) and legal action. This approach will improve the quality of the initial bid (in its relevance and fitness for purpose, rather than the cheapest). It may ensure local support of supply, improve proactive maintenance and, incredibly, make the whole procurement process a great deal cheaper.

A model of procurement

The ideal set of procurement documentation should universally guarantee the deployment of quality energy solutions, regardless of the supplier. How do we go about achieving this?

- Place the emphasis on the performance of the installed services.
- This service reliability will be the sole responsibility of the supplier; after all, it is what they are selling. Therefore, allow them to propose a solution against the clearly stated functional requirements of the customer.
- Assert a preference for local suppliers and those who are experienced in PV design (compared with those who simply source and consolidate parts).
- Recognise companies with a positive track record, and challenge those who have supplied inadequate equipment in the past.
- Itemise all of the important aspects of the supply. 'Comply', 'non-comply' or 'part-comply' responses are required from the potential supplier to each item.

The documents belong to the organisation purchasing the solutions – the Uganda MoES in our example. Guidance notes are supplied for the purchaser to manage the procurement process; for example, in how to address part-comply and non-comply statements. No external or detailed PV system knowledge is required to use them. Supplier selection criteria will be included and the documents form part of a legally binding contract with the resulting provider. This structure is successfully applied in hundreds of commercial and industrial procurement processes.

The value of the procurement specification is broad. The above list addresses the project process, but the devil is in the detail when it comes to ensuring quality. Choosing stainless steel nuts and bolts over cheap steel versions spells the difference between months and years of corrosion-free use; between replacing parts with ease in the middle of nowhere and needing special cutting equipment. Fitting plastic washers can prevent damage from lightning; cables can resist years of environmental exposure; and labels can remain readable for years on end, if they are UV resistant and laminated. Most of these measures add minimal cost at the procurement stage, but they save huge amounts of time, money and effort later on. They all contribute to lasting value.

Procurement should stipulate training for any individual within (or nominated by) a government who purchases PV solutions. This is because standard procurement processes may well entail activity across different government departments or agencies: for example, the Ministry of Education buying computers, for which the Ministry of Energy is – sorry – charged to find a power solution. If the former buys the cheapest option, it is likely the latter will have to spend disproportionately high amounts. This needs inter-departmental agreement and a trained consensus on quality and value.

As for suppliers, many PV systems are supplied by companies for whom PV is not their core business (for example, air-conditioner companies or automotive suppliers). Although the electrical and mechanical requirements of PV systems may *appear* the same as other electrical equipment, they are actually quite different and in many cases unique. As I mentioned earlier, some suppliers might appreciate a quality solution when they see one, but will nevertheless offer the lowest bid to win business. Others simply do not understand the finer requirements of professional PV systems.

In a nod to our previous chapter on knowledge sharing, training private sector bidders can help to draw a line underneath the past experiences of poorly supplied systems, and subsequent effect on reputations. It can be conducted for individuals or groups and can be scheduled in manageable segments to ensure minimum disruption to normal business operations. Typical training modules would ideally cover the following: principles of cost optimisation; understanding (and compliance with) the technical specification, supplier criteria, and international standards; and additional areas of value – items which often cost very little if considered at the design stage (for example, vandal resistance, remote monitoring and anti-theft design).

Procurement and value

Approaching procurement specifications in the way I have described will result in high quality and high value energy services for the end user. The specification process will be more expensive to implement, but the overall project budget will often be far lower, and the implemented solutions will be superior in several ways. Operational costs for the supplier will be lower, the system can be maintained more easily and, most importantly, all stakeholders can have confidence in the energy services being provided. This is where the true value of PV systems really lies.

Computers in a classroom can be integrated into the learning curriculum with the knowledge they will operate for the entire school life of many students. Medical refrigerators can store vaccines without risking damage from lack of power. Businesses may communicate without interruption throughout the working day. And, probably most importantly for many local people, the entire sporting season can be watched on TV.

As far as equipment is concerned, the specifications will *optimise the cost* of PV solutions. Reliability emerges here, not only from the quality of the design and components, but also from a commitment to support from a local supplier. When designed correctly, the maintenance is straightforward and can be conducted by individuals with modest training. The equipment is safer (electrically, mechanically and chemically speaking) for installers and users alike. The environmental impact, from installation to the end of the system's life, is significantly reduced.

Structured procurement will make a huge positive difference to the large, bespoke installations for telecoms, schools, hospitals and community energy services (mini-grids in particular). They will also make infrastructure networks of small systems (the regional 'solutions in a box' needed for Wi-Fi coverage, monitoring applications and transport support) far more valuable, reliable and deployable.

Anybody sensing a 'but' at the end of this statement may congratulate themselves. These specifications will *not* impact quality and reliability in one of the largest and most critical areas for energy services: consumer applications. For personal phone charging, solar home systems and service-specific products – TVs, radios, fridges, fans, computers – a different approach is desperately needed.

CHAPTER 13

Why creating an energy services brand is essential

Abstract

Brand is the mechanism that can perform the task of procurement standard for the general population wanting to purchase solar PV personal and consumer products. Brand also enables a long-term, commercially sustainable and broad product portfolio.

The most powerful motivation to establish a brand is that, quite simply, it is essential to profits and ongoing commercial success.

The right products will not be the cheapest, so they need to be both recognised for their value and distinguishable from the spurious performance claims of many other products.

The right products need brand identity, and brand identity, in conjunction with product quality and carefully developed media, is needed to represent and reflect reliability and dependability.

The brand needs to build confidence for other applications, larger and longer-term projects. This launch pad gives us the means to develop towards 'guaranteed energy' for bigger systems – for infrastructure, policy and investment.

The PV industry is over 50 years old. Until a few years ago, it was consistently the fastest growing technology sector in the world, with an average CAGR (compound annual growth rate) of over 30% for 30 years. Its annual turnover exceeds $100 billion; it has thousands of component suppliers, hundreds of manufacturers and dozens of multi-billion-dollar turnover corporations. Yet, apart from a few regionally familiar names and logos in particular sectors, there is *no global brand*. None. Most of the general public would be unable to identify a particular PV manufacturer or supplier. Stop for a moment, and try to name another mature sector this huge or so commercially significant that dares to function without recognisable branding or positioning. Why is this?

We don't know, or indeed care, who makes the engines in our vehicles, or the batteries in our smartphones. These items are essential, and in many ways define the equipment they enable. But these technologies are useless until they are integrated into a system. Similarly, PV *only generates electricity*; as consumers, we have no use for this in its own right. We require the benefit of this electricity in the form of a service: communication, light, entertainment; or simply a cold drink. PV cells generate at low voltage, so they must be connected in series. The resulting modules generate more 'useful' levels of energy,

http://dx.doi.org/10.3362/9781788530705.013

but only during the day, and only when suitably located outside. All-in-all, not very convenient.

PV electricity must either be used at the moment of generation, or connected to other equipment to use it at another time. For grid-connected systems, this requires an inverter; for stand-alone applications, a battery. So consider the products which emerge at this stage. Inverters, for example, turn PV energy into AC electricity at a specific voltage and frequency. Nobody could care less about this. Batteries store DC electricity at a given voltage – again, of no practical interest.

Some brands within the industry have attempted to sow their names into the popular consciousness by advertising at football clubs or sponsoring Formula 1 teams. Overall, these efforts have had short-term influence on buyer decisions. For a PV manufacturer to achieve popular recognition would require huge ongoing financial and creative support. There are over a thousand battery manufacturers in the world. How many can you name? The Duracells of the sector spend tens of millions on PR and marketing. Yet, if you needed batteries for your torch and the shop only stocked Energizer, would you go elsewhere? If you decided on a particular computer specification (processor speed, memory capacity, screen size, battery life and so on), and you found exactly what you were looking for, would you refrain from purchasing it because it didn't have an 'Intel inside'?

Engines, batteries and PV modules: these technologies
are useless until they are integrated into a system

For practical reasons, it is difficult to display branding on PV components. While PV cells do exist with tiny logos on them (similarly there are modules with company colours and names), their only visible surface is optimised for generating electricity. Displaying a logo here will only block light and compromise its function. Realistically, branding is relegated to the side or back.

As it goes, the industrial off-grid PV industry was starting to exhibit something close to brand recognition in the 1970s and 1980s, when PV modules were often assembled by hand and boasted varying reliability. Mass-manufacture by companies such as Solarex and Siemens became sought after (particularly within the oil and gas sector, where PV modules were specified by name within procurement documents).

After this, the 'grid-connect' markets became dominant. 'Off-grid modules' could not compete commercially with larger, higher voltage grid versions. And because the cost of the PV component was so critical for grid-connect (at the time over 70% of system cost versus around 25% for off-grid), modules became a commodity bought almost exclusively on cost. Since the beginning of the millennium, grid-connect applications have accounted for over 80% of PV module installations (rising to more like 90% in the last decade). Manufacturers have come and gone, technology has advanced and prices have fallen; but every year there is a different top 10 largest manufacturer list.

In such a market, the challenge of establishing a secure foothold for a brand becomes self-evident.

The breakthrough brand?

I predict that the first brand that utilises PV will *not* come from the grid-connect market, where 90% of PV is presently installed.

When connected to the grid, PV modules simply produce electricity. Their only tangible value lies in saving ongoing mains electricity costs; and to a secondary extent, generating clean energy. Reducing it to basics, it does 'the same thing as before, only better'. Brand value requires more than this. By contrast, autonomous applications, particularly in the developing world, change people's lives; immediately and visibly. PV enables light, communication, refrigeration, entertainment, education, health and so on.

The first truly global, truly recognisable brand will come from the so-called off-grid sector. And it will likely come from the developing world. When I first considered this, I tried to research examples of brands which have originated in poorer countries and grown to prominence in richer ones. I couldn't find any.[1] While I'll decline to go into issues of globalisation or the specifics of brand theory, let me assert why PV is sufficiently special, and why its global prominence is likely to originate from these developing world, autonomous energy markets.

> The first truly global, truly recognisable brand
> will likely come from the developing world

Tangible personal value

Stand-alone PV, with the exception of some larger bespoke installations, is situated close to its beneficiary. There is an obvious visible connection between the PV panel in the sun and the subsequent service it enables. Once again, mobile phone charging acts as a perfect example. Plug a PV product into a phone; the phone shows that it is charging. PV lighting products also display a direct and immediate benefit for the owner.

Historically, autonomous PV has been under-utilised – as a power supply, or worse, as a battery charger. Even for expensive modern-day solar bags and folding chargers, its value has been both ill-perceived and unfulfilled. But we are now witnessing a convergence of enabling factors: widespread demand, maturing technologies, a better understanding of electricity (as opposed to 'energy') and the value of its reliability. These will allow stand-alone PV to enjoy a direct association with the service it is enabling. PV can and will provide communication, education, health, entertainment and, yes, that cold drink.

Contrast this tangible, hold-it-in-your-hands value with a grid-connect installation. Here, PV produces electricity only during daylight hours and makes no difference to electrical services. Furthermore, the PV system shuts down if the grid is not present; it serves no backup purpose. Silently and obediently, it feeds a detached and inscrutable beast with energy, displaying no direct consequence of its operation: in other words, no value. Grid-connected PV competes with every other form of electricity generation out there: fossil fuels, nuclear and renewables. In this arena, we see again the consumer obsession with cost, moaning about the financial burden of electricity, let alone the unfair subsidies they are forced to pay for 'green' sources. The manner of grid connection fosters this very entitlement culture – 'I deserve and demand more, better and for less' – which bays for almost infinite power to within a few metres of their armchairs. And screw the environment while we're about it.

Witness instead the excitement of people in rural areas in energy-poor countries, when they see and experience what electricity can do; witness the revelatory moment when they learn that they can *own* their electricity generator – and carry it with them or install it on their house. They treat products with great care, once shown how; and become the envy of their neighbours.

We are now witnessing a convergence of enabling factors: widespread demand, maturing technologies, a better understanding of electricity and the value of its reliability

---

Riches off the grid: why we need brands

Autonomous PV increases in value with its distance from reliable mains electricity. Also you are going to get the most out of the generic technology in sunny areas – especially in places with year-round good levels of light. Look at a global map of sunlight intensity, then compare it against a 'lights at night' map from NASA (which shows the global spread of mains electricity grids, and, more revealingly, where they are lacking). You will see that one is pretty much the inverse of the other. Easily identifiable, sun-drenched land masses (most of Africa, India, Central America, a good chunk of China) have either no or unreliable mains electricity – or at least limited distributed alternatives. This means that PV has greater intrinsic value.

Rural, un-electrified areas in the developing world: this is where PV is most useful, has the least competition, is appreciated the most, and has the strongest and most obvious association with the service it enables. As an aside, it is also where PV – beyond the capabilities of any other technology – has the potential to help in the fight against poverty and generate social-economic opportunities for the poorest billions of our kind.

All of this can be viewed as opportunity. However, the most powerful motivation to establish a brand is that, quite simply, it is essential to profits and ongoing commercial success. More and more people and organisations are recognising the need for electricity in energy-poor regions and becoming aware of the commercial potential for PV.

But if you analyse the present and historic sector, it raises some doubts about the reality of sustaining a business. Over the last 10 years or so, a number of high-profile companies have entered the 'off-grid PV market'. Some have been purely for-profit, some have labelled themselves social enterprises, non-profits or similar. Most have had initial capital – either straight investment or 'philanthropic donations', and most have produced good (or at least good *enough*) products.

Marketing and PR (some with an emphasis on product value, others 'helping the poor') have generated sales and real-life profits. But no company has sustained significant volume or profits for more than a few years. The most successful suppliers now are different from three years ago, and those different again from their predecessors.

There are many reasons why companies fail to reach their promised or expected volumes. As more competition enters the market (with most products now being made in China, and with cost the stated priority), many suffer product quality issues – a familiar and widespread cycle in recent years. Some adopt business models which are not transferrable; for example, the high population density of Bangladesh supports marketing and sales to volumes of customers which are unrealistic in distributed African communities.

Others have been very closely associated with a particular ethnicity, tailoring their sales literature and product design accordingly. Some have competed primarily on cost, with limited margins. Some have been funded with restrictions on geographical operations (USA funding for Christian areas cannot typically be used for Muslim ones.) Organisations such as the World Bank are generally funded either on a country-by-country basis, or by countries of a common language – thereby limiting half of their name's function.

Branding is only part of the solution, and is by no means a guarantee of success in itself. d.light is a good example of a company who produced a solar lamp, branded it and very effectively spread their name through customer and donor markets. But even after tens of millions of dollars and with early-mover advantage in the 'new generation' solar lighting market, they have failed to reach even their own specified level of impact.

But in conjunction, brand, product development and visual media are essential to seed longer-term success. All require high capital funding and sustained support and development. And that kind of backing only comes from commercial companies who can justify expenditure against projected earnings. Or, for that matter, against predicted risk.

Branding and value

We saw earlier that a valuable electrical service is, before all other things, a reliable one. This demands that we design products or systems for the target users and the conditions (particularly the climate) in which they are used. Branding plays a crucial role here too, far beyond a matter of image and marketing. The use of media is critical in maximising the likelihood of correct use and care of the product. The successful brand is one which promotes understanding of quality and service: a message that the brand cares about energy-poor users as active participants and agents of reliability; that they are the priority, not secondary, passive 'consumers' of difficult-to-use products with foreign instructions. But most of all, the brand needs to reduce the emphasis on product cost and distinguish a product range from the oceans of cheap rubbish available on the market.

> The successful brand is one which facilitates an understanding of quality and service: the message that the brand cares about the energy-poor user

Product performance claims

The cost of a stand-alone PV product is directly related to the energy it can produce; the PV component cost and size are major factors, as is the battery capacity and technology. Less PV or a smaller battery means a lower product cost, but worse performance.

Customers do tend to overlook this. PV products such as phone chargers, LED lamps and battery chargers look very similar, often using the same crystalline PV technology. And, as we know, in the absence of any other way to compare products, customers, distributors, NGOs and governments are more likely to believe rather than question the inflated claims of a product's performance, and will default to the lowest cost.

Product performance claims, often written in English or using simple numbers and pictures (hours in the sun equating to hours of light produced at night, for example), often stretch the realms of what is reasonable. Elsewhere they are simply false. A product can claim 'up to 6 hours of LED light from 8 hours of sunshine'. But this claim can only be met if the product battery is brand new, being used at or below 25 degrees centigrade and 100% charged with the LED light on the lowest of four brightness settings. To add insult, the small print can state '8 hours of full intensity sunlight' or similar – which means two or more days of cumulative daylight exposure, even in somewhere sunny like central Africa.

This is doubly misleading. Its disappointing performance does nothing for the reputation of PV as a generic technology. Neither does it inform customers of the value potential – the fact that a well-designed, well-supported product might be the most expensive option on the shelf.

Branding – particularly branding which reflects the true potential of the product – is the only mechanism to overcome this in a sustained manner, for a product range that needs to encompass everything from personal devices to national infrastructure.

The service and the brand

Stand-alone PV has a direct and visible association with the services it enables. Those who need them can own their PV products and systems. The right products will not be the cheapest, so they need to be both recognised for their value and distinguishable from the spurious performance claims of other products. The right products need to inspire confidence in the generic PV technology; but furthermore, like the most recognisable global brands (such as Apple, BMW, Omega), they must embody the services the user wants.

The right products need brand identity, and brand identity, in conjunction with product quality and carefully developed media, is needed to represent and reflect reliability and dependability (the present barriers to big brand names 'going solar'). The brand needs to build confidence for other applications, larger and longer-term projects. This launch pad gives us the means to develop towards 'guaranteed energy' for bigger systems – for infrastructure, policy and investment.

Endnote

1. I'm sure there are examples within specific industries that their members can identify – maybe tea, coffee or fabrics – but I don't know of any in mainstream society.

PART FOUR

People are the cause of poverty; and the solution

CHAPTER 14
Why poverty persists

Abstract

Before entering into a specific discussion on energy, I feel it is important to mention at least some of the broader factors which define the fate of so many humans.

First, a critical reality: Global poverty and inequality are not due to a lack of money, limitations in technology or the scale and diversity of developing world populations. Global poverty persists because sufficiently large numbers of people are not motivated to address it.

In this chapter I describe some of my experiences and observations regarding people; *and the manner in which inequality seems to be regarded within society. These observations have a bearing on the potential for addressing the issues of poverty in general. At the very least they will help set the scene for the discussions on energy services which follow.*

Summer 1995. It is the end of the working day, with a deep orange sky steadily fading to the right. It is the busiest road in the country – stretching to the capital some 300 km north. Most of the heaviest traffic is headed towards me, out of Blantyre, the second biggest city. This is Malawi in central Africa, one of the poorest countries in the world.

Men, women, boys and girls. Tall and short. Some men wear suits; others are barely covered by tattered, ill-fitting clothes. Some are slim and muscular, some slim and frail, carrying hand-fashioned walking sticks. The women steal the show wearing bright greens, yellows and blues, their long dresses and headgear shiny in the dusty and dim surroundings. I am the only white person, a novelty to many locals.

Oncoming vehicles, many long past their retirement, rattle and screech as they struggle along. Buses with arms leaning out of all the open windows, cars of all ages. The engines of the older, typically overladen, lorries cough and stutter with the slightest acceleration or incline. The gearboxes crunch, the exhaust pipes crack against the bodywork. A backfire startles everyone within earshot. These people have evidently lived through conflict.

Minibuses are seemingly everywhere, ferrying commuters between the urban and more rural areas. The drivers are pushing the limits of their vehicles, scattering the roadside walkers. Amid our vigilance, flies harass and distract; while mosquitoes pose an altogether stealthier, covert threat.

As the light fades on this particular night, animals pulling carts are being directed by thin men and boys, some of whom are barefoot. As more people join the exodus the human scent starts to develop, the periodic smell of the animals mixed with exhaust smoke creates a disturbingly powerful cocktail. Some young men have personal radios held against the ear. Local stations playing bouncy and melodic, definitively 'African' music.

http://dx.doi.org/10.3362/9781788530705.014

Numerous women are carrying goods on their heads. A large bunch of bananas, a heavy bag of rice, a woven basket full of something covered with cloth. They maintain their elegant progress with arms by their sides, only using a hand to steady the load above when an approaching vehicle demands they quickly alter their course. Many have babies strapped to their backs. As soon as these babies are able to walk, they will have to, braving the roadsides with the grownups as vehicles speed by within an arm's length.

Older children, walking many kilometres from school, are talking, laughing and messing about as if on a safe urban sidewalk. Many set out on this pilgrimage at dawn. They stand out in their white uniforms – probably the only one they own. They will have to wash it, ready for tomorrow, should they falter on the smoky, hazardous walkways, as darkness approaches.

Now it is night, and the scene has turned a sinister combination of dim, obscured vehicle lights, dark skinned pedestrians with faded or dirty clothes, and a shadowy, intractable roadway bereft of street lights or road markings. An approaching lorry induces a sudden panic as it lunges to avoid a concealed pothole. The vibration peaks as it passes without incident. A brief, still moment of relief makes way for a cloud of abrasive dust and toxic fumes, which engulfs all around.

Here, people leave the roadside procession, along informal, unmarked, dusty single track walkways, leading into darkness.

So far in this book, I have not talked about poverty in detail. This may strike you as odd in a book concerned with energy in the developing world. The introduction piece to this chapter is a living snapshot describing my first exposure to the realities of daily life in a developing nation. It may be familiar to you, or alternatively challenge the instinctive image of 'Africa' as a barren wretched landscape which you have received from the media. You will notice that it describes people on the move; a land bustling with purpose. It shows a civilisation alive with infrastructure, rules, customs and collective aspirations; upwardly mobile, trying to take care of business. Noticeably, and in deliberate contrast to the familiar images of suffering in countless aid campaigns, it is a depiction of everyday life, as opposed to everyday death. Nevertheless, this progress, and the potential it signifies, continues unlit, in a shadowy facsimile of our own secure and comfortable existence.

Poverty is a devastatingly emotive subject, which bewilders, shocks and shames us in equal measure. Its existence in a wealthy world is an assault on the fragile collective conscience, confounding our ability to reflect and analyse; and dividing the opinions of those who seek to change the world for the better. Broadly, it is a subject vulnerable to misinterpretation; a mixture of immediate emergency responses and robust longer-term considerations. Specifically, when it comes to providing energy services to the developing world, the emotive rhetoric of poverty relief proves more of a distraction than a motivator. And motivation, as we shall see, underpins what follows.

In this area, I, like many, consider the short-term, palliative focus of 'poverty alleviation' to be a mistake. I believe that the route to solving energy poverty is to engage the commercial markets of the world, furnishing them with the opportunity to sell energy services to consumers who demand and can pay for reliable solutions. I hope to demonstrate that by ensuring quality and versatility, and by using innovative ways of selling to those already spending (by this I mean governments, organisations, families and individuals), we can make sufficiently strong profits for commercial growth. If we achieve this primary goal, we will have the financial and logistical means to expand on a massive scale, increasing access to quality energy services for the poor.

It is now widely understood that access to modern energy services is essential in the fight against poverty, and in the pursuit of social and economic development. But before entering into a specific discussion on energy, I feel it is important to mention at least some of the broader factors which define the fate of so many humans.

It is now widely understood that access to modern energy services is essential to poverty alleviation and social and economic development

---

People: a perspective on poverty

Dependency on fuel and aid; enforced political allegiances; trade restrictions imposed by the rich; climate change; the consumer practices of rich nations (particularly supermarkets); and the arms trade. There is strong evidence that all of these are keeping entire nations and continents in poverty and restricting development. However, I am not going to claim to have the last word on what poverty is, how it came to be, or what causes nations to become trapped in the cycle of poverty. As a subject, it is vast, complex and contentious. Instead, and this is an exercise in perspective, I want to describe some of my experiences and observations regarding *people*; and the manner in which inequality seems to be regarded within society. These observations have a bearing on the potential for addressing the issues of poverty in general. At the very least they will help set the scene for the discussions on energy services which follow.

First, a critical reality: Global poverty and inequality are not due to a lack of money, limitations in technology or the scale and diversity of developing world populations. I shall discount these factors in their turn. Rather, global poverty persists because sufficiently large numbers of people are not motivated to address it.

Much like our received picture of 'Africa' in the mind's eye, the term 'poverty alleviation' is much generalised. Some think of it as meaning aid, or charity donations, while others regard it as a programme of social and economic development. It is not always clear which of these definitions has been adopted by a given charity or non-profit organisation (or even some 'developmental' organisations). Many – Oxfam and the Red Cross to name just two – are engaged in both charity and development work.

There is a big difference between charity and development. Charity is a handout, a momentary gesture. In certain circumstances, such as disaster and emergency response, it can be vital; but 'development' – helping people to help themselves into the future – is what most disadvantaged people really need. The vast majority of both charity and development activity is within poor countries and regions. Unfortunately, this confusing terminology is representative of the whole sector. The developing world is a mess of hypocrisy, counterproductive initiatives and misplaced priorities. My conviction that energy poverty needs to be solved by commercial market mechanisms is based on rational logic, but it first materialised as a result of my losing confidence in the whole charitable and development sectors.

The development sector

The development sector is complicated and frustrating, and at times difficult to fathom. I have experienced several specific frustrations, looked in the eye by a client with a development project and told: 'your proposal is clearly the best, the cheapest and highest value, but the contract has to go to another bidder. Sorry.' (Asking for an actual reason prompts a performance akin to a teenager trying to get out of housework.) My experiences of hidden agendas and unethical priorities are echoed by anecdotal information and conversations – hundreds of them over an 18-year period – with those at the hard end of the problems. I cannot articulate why, but it is clear that even the best products, the most capable organisations and the highest value propositions are no guarantee of success.

Meanwhile, disagreement continues regarding aid. When one globally respected economist (Jeffrey Sachs for example; Sachs, 2005) makes the impassioned case for increased aid spending, while another (e.g. Dambisa Moyo, 2009) claims that it is damaging, the discussion is clearly more complicated than most people can understand.

Global poverty and inequality are not due to a lack of money, limitations in technology or the scale of developing world populations

---

But my issues with the sector are about more than just money. I strongly question the day-to-day intentions of the hundreds of organisations involved in 'poverty alleviation'. Without doubt, there are truly committed individuals and honourable intentions. But the non-profit sector is a world of contradictory objectives. There are several official definitions of poverty within the international community. Statistics get thrown around, based on assumptions of such breadth that they may as well describe commodities rather than suffering people. The international agencies, officially tasked and funded by us in industrialised nations to address inequality and injustice around the world, seem to be more like political parties or self-interested banks than anything else. The sheer weight of books, papers and initiatives, and the attendant circus of personalities is an industry in its own right, and can make one call into question the validity of the whole movement.

Poverty is obviously a complex tragedy, but the organisations tasked with solving it often seem to make the situation far more complicated than it needs to be. Good intentions, empathy, missionary work, aid and philanthropy are all very well, but, ultimately, they have not delivered (and I believe *will* not deliver) what is needed in the long term. My specific field of energy, for example, needs large and capable commercial suppliers with a primary focus on the developing world because, quite frankly, the past and present methods of addressing energy poverty have not worked.

Quantifying poverty

The most common measure of poverty is of a daily income in a currency we can understand; such as those living on less than $2 per day. However, there are problems, taste notwithstanding, associated with trying to put a universal dollar figure against poor people's circumstances. Other assessments look at hunger or access to basic human needs, the definition of which also varies: food of a minimal nutritional value, water, basic health and education. Even electricity has been called a basic human right; though as we have seen, this misses the point – the critical metric should not be energy or electricity, but the services that derive from them.

Elsewhere, there are attempts to frame poverty as an index of suffering, including social exclusion, fear, destitution and those who have lost hope; those unable to change their situation without the help of others; and those vulnerable to risk. Here, depressingly, we find the human-induced miseries, not confined to war and oppression, but of trade prices and policies, and the low-cost manufacturing demands of the industrialised world, separating families and subjecting workers to multiple health and safety risks.

Definitions can be useful, but also obstructive. What are the common factors for those in poverty, or close to it? The circumstances vary, as do locations. These include low agriculture potential, low rainfall, high altitude, a shortage of employment options (in agriculture or alternatives), and those struck by natural disasters such as floods, earthquakes, volcanic eruptions and severe storms – often struggling many years after such events.

There are poor people living in towns and cities, in both rich and poor countries. But the majority of the very poorest people live in rural areas. Broadly (and this will become relevant in the context of energy), the major categories of poor people are subsistence, or near-subsistence, farmers. Using some representative figures, we find that half the world's hungry people are from smallholder farming communities; with a further 25% in herding, fish and forestry or landless; and the final 25% in urban areas.[1] Large numbers of poor define their regions and their countries. In Africa's case, the entire continent is synonymous with poverty.

Background to poverty

What makes countries rich or poor? Globalisation, corruption, natural disasters, tribal feuds and a nation's ability to meet the environmental challenges all vie as candidates, with many diverse opinions as to the role played by each of them. There is also a growing understanding of human psychology on the subject; of why people who could help fail to do so; why even those who were once impoverished distance themselves from their still-struggling communities. Furthermore, when we attempt to look at the longer game, the slower evolution of peoples and nations, we find that the underlying economic and institutional reasons for affluence or poverty take centre-stage.[2]

On the country level, economists have different views regarding the relative importance of the causes and conditions that make countries richer or poorer. However, 'good institutions' are clearly a primary factor.

> Among the good economic institutions that motivate people to become productive are the protection of their private property rights, predictable enforcement of their contracts, opportunities to invest and retain control of their money, control of inflation and open exchange of currency. (Diamond, 2012).

Acemoglu and Robinson (2013) maintain that history also plays a key role, and it varies from country to country. This includes the historical durations of centralised governments, linked in their turn to the durations and productivity of farming. Then we see the effects of colonisation, whereby ruling powers, or 'extractive' economic institutions, used forced labour and confiscation of produce or institutional incentives to reward work, depending on the natural resources of the colonised nation. (Termed 'the reversal of fortune' as colonies originally richer and more advanced tend paradoxically to be poorer today; and vice versa (Acemoglu and Robinson, 2013: 79-83).

Next is the 'curse of natural resources', so named because of the 'many ways in which national dependence on some natural resources tends to promote corruption, civil wars, inflation and neglect of education' (Diamond, 2012). Dependency on overseas markets and the global economy are also important factors.

The economic challenges facing a given region are largely defined by their geography; in the difference, say, between a tropical and a temperate climate. Tropical countries suffer economically. Diseases, with no cold winter to kill them off, affect the productivity of large sections of the working population. Women, meanwhile, are fully engaged in caring for babies – the family 'insurance against the expected deaths of their older children from malaria' (Diamond, 2012).

Agriculture is the life support for the majority of the world's poor, but they have to work far harder for less, in large part due to the climate. Tropical plants store less energy in the parts edible to humans, pests and insects cause more disease in crops in tropical regions, the soil typically has lower nutrient levels, rainfall is higher on average (which dilutes nutrients and can damage crops directly), while the higher temperatures in tropical climates reduce the organic matter falling to the ground from being incorporated into the nutrient cycle.

Landlocked countries have the additional burden of high transport costs and dependency on their neighbours – cargoes being significantly more expensive to transport by land than by sea. Low population densities also raise transport costs. This discourages commercial development as well as NGOs and government organisations. These populations are where the value of PV is highest.

> We all know, from our personal experience, that there isn't one simple answer to the question why each of us becomes richer or poorer: it depends on inheritance, education, ambition, talent, health, personal connections, opportunities, and luck, just to mention some factors. Hence we shouldn't be surprised that the question of why whole societies become richer or poorer also cannot be given one simple answer. (Diamond, 2012)

Why poverty persists

Poverty propagates for many reasons. To understand fully why it persists within a supposedly advanced civilisation requires many skill sets and a depth of understanding in diverse fields. Economics, politics, energy, technology, transport, climate science, agriculture, flora and fauna, medicine, disease, language, cultural practices, religious observation and history (and this is only the beginning) all play their part. It won't escape you that expertise in the whole list is clearly beyond the strategic capability of even the most gifted of individuals. But organisations, for all their collective expertise, also struggle to grasp and consolidate these factors influencing poverty, let alone understand the resulting problems or their solutions.

Numbers that don't add up

Large international bodies such as the World Bank, in their efforts to understand poverty, face a seemingly insurmountable challenge in the form of statistics. Their frustrating variability and sheer scale defy consensus, and straightjackets an organisation which requires some sort of empirical foundation upon which to strategise and act.

Elsewhere, smaller agencies may hold limited and localised data; but this detail loses relevance if you attempt to apply it to other regions – realistically, you cannot generalise about problems of this scale. For example, the kinds of daily trials experienced by high altitude communities in Nepal will not compare meaningfully with those of fishing villages in Malawi. The challenges facing these people are both similar and worlds apart at the same time. Separated by geography, environment, politics, culture and more, many of the inhabitants of these two places can technically be said to live 'in poverty'. But this attempt to unite their humanity and condition ironically only serves to conflate them, inaccurately, as statistics in a larger, intractable global dilemma, where the numbers fail to add up to a meaningful answer.

In terms of our collective global capabilities, there are no barriers to ending poverty. International organisations, with a direct mandate, have certainly had enough money and resources to hand; but after at least three decades of purportedly focused effort, we remain in a world of grotesque inequality. Amid the opinion, analyses and finger pointing, several fundamentals are clear. When we look at poverty from a global perspective, it is people (and more specifically, their motivations) who prove the common factor. The movement of money, while central to promoting social and economic development, proves critical to how people are motivated in general. The problem, it seems, lies not with a lack of money, but with our preoccupation with it.

A preoccupation with money

International organisations call for billions of required spending to end extreme poverty. Global movements are created to raise funds, either for specific causes (such as emergency relief, Malaria or AIDS) or in a more generic campaign against 'poverty' as a whole. Charities and non-profit organisations of all sizes make appeals for personal spending for a specific cause. In thousands of studies, economists analyse the financial mechanisms of the developing world in an attempt to make sense of the movement of money; how much is being spent, by whom, why, where and on what.

But there is a difference between money as a numerical, statistical indicator of wealth and money as an enabler. The former is an abstract; the latter requires effort. Elsewhere in this book, I have made a distinction between 'energy' in its pure form and *what we do with it* – the quality of the lifestyle services it enables. Here again, a vast monetary resource is dependent upon humans, their attitudes towards it, their instincts, aspirations and assumptions; and their ability to utilise it.

I could explain, at length, how the cost of *this* essential energy service can be met by *this* small budget. Or I could quote detailed programmes raising *this* number of millions in order to eradicate *this* disease; providing computers to each child if we raise *this* amount of money within *this* time frame. But it would all be a distraction. The stark reality to accept is that *we* are the problem; relatively wealthy people, with plenty of money at our disposal. Which we devote, with an increasing and oddly pathological ingenuity, to ourselves. Ultimately, I consider a focus on the commodity of money itself to be misguided. The real issue is how we choose to spend it.

Opulence and charity

In January 2017 Oxfam reported on the magnitude of wealth inequality in our world (Hardoon, 2017). It warned that the richest eight men across the globe control a combined wealth of \$426 bn (£350 bn), as much as the poorest 3.6 billion (half) of the world's population.

This is the world we live in. A world whose markets have enriched companies and individuals alike with vast fortunes, which far outstrip the gross domestic product of nations. Where exclusive services provide a means for the super-rich to dispose of their cash, extending the limits of opulence beyond the limits of 'taste'. (Ignoring, if briefly, that word's context against a backdrop of half of the world's people struggling to make just a couple of dollars per day.) A concrete Aalto doorstep (a small block of concrete) costing $3,500; a gold backpack ($1,650); a diamond encrusted Bluetooth headset ($50,000); a Tiffany tennis ball container ($1,500) (Cool Material, n.d.). Stepping aside from this game show gone wrong, compare for a moment the wealth of generic industries: a US$265 bn personal care and beauty market (Lucintel, 2012), $1.088 trillion on tourism in 2012 (Mohn, 2013), billions more on designer clothing; on beauty treatments, relaxation therapies, and so on.

By contrast, the amounts of money required to make radical positive impact are tiny compared with the disposable incomes of hundreds of millions of people. Inspired campaigns endeavour to raise funds by stating minimum amounts of money required to make practical changes to an individual. For example, a donation of $3 pays for a vaccine for a child; $300 provides shelter for someone who has lost their home; $30 per month provides education, food, water and healthcare for a child; $50 provides clean drinking water for an entire family for a whole year. Most of us have seen these adverts on TV and in the press since we were children; their production style and narrative all-too-familiar in a media-saturated age.

The amounts of money required to make radical
positive impact are tiny compared with the disposable
incomes of hundreds of millions of people

Considering these tiny amounts of money and the dramatic impact they can have, against a snapshot of the grotesque amounts spent on non-essentials, it would seem an easy step to make this impact real and lasting. But the fact of the matter is obvious: zooming out from the actions of well-meaning individuals, collectively we don't make these donations.

Coercion, choice and change

What is my point here? Firstly, for the record, I am saying that it is a great shame that people with such extreme wealth, and those who have significant disposable income, are so inclined to spend so much on themselves and so little on others; that humans as a species are not driven to take care of our own.

But this is not a soapbox. There is no clear line between quality, luxury, opulence, greed and stupidity. And as we know, perceptions of wealth, as soon as a fraction of it becomes 'disposable', are relative. Ultimately, a person's money is realistically theirs to spend and I am not about to dictate how they should go about it. Poverty is not going to end overnight just because we are all told to do so: people feel aggrieved if they are instructed to provide money to causes they have not chosen themselves. Forced charity is not charity, and it seeds resentment. On the contrary; poverty will end out of choice, by people who are responsible owners of their actions.

It pays to step back from the detail at this point, and consider the global reality. While it is simplistic to say 'we have enough money', given the generic complications in applying money to global problems, you cannot escape the fact that humankind is capable of removing at a stroke the poverty and the suffering of the innocent from this world.

Poverty will end out of choice, by people
who are responsible owners of their actions

Consider some of the most celebrated human achievements in the modern age. The ability to escape the Earth's gravity and land on the moon (half a century ago, let us remember); the mapping of the human genome and its implications for medicine; the fabrication of billions of switches into a single tiny microprocessor, revolutionising information and communications. Any such undertakings successfully apply huge funds and the joint efforts of individuals to a common purpose; yet all these achievements are against a backdrop of suffering. This is how the world functions. We cannot fix everything against a backdrop of how the world *should* be. Most efforts to understand and eradicate poverty are preoccupied with the immediate need for money – aid, donations, sponsoring a child, international support and so on. This is understandable, but to a large extent it is a distraction from another, starker issue. The magnitude of inequality can provide a clue to what this is. The money exists, for certain, in the coffers of nations, in defence budgets, property, hedge funds, housing bubbles and bonuses; and sometimes in our pockets. But a significant characteristic of our nature is devoted to keeping it there, and the motivation to change this state of affairs, or look further than our immediate circle to spread the wealth, does not come naturally to us.

Affluence in a poor land

Global inequality demonstrates that it is people who define global circumstances. But there is more to this than merely an imbalance in wealth or possessions. There is a depressing trend of opulence literally in the face of extreme poverty; perhaps exacerbated by modern media and the impression of a shrinking 'global village'. This is nothing new. The world's poor have always existed in close proximity with privilege and wealth, more so since industrialisation and urbanisation. More puzzling is the increasingly common flamboyant display of spending by those arguably capable of the greatest empathy with the suffering of our world; put another way, the attitudes of people who used to be poor, towards those who remain so.

A Chinese friend of mine grew up in a home without electricity or water and had to walk outside to a 'horrid' toilet. The family had to walk to collect water, fuel for cooking and generally struggle to live. My friend would have been the target of 'aid' and rural development funds. Nowadays hardworking but well-off, they spend a great deal of money on their favourite pastime of shopping. Designer bags, expensive earrings and bracelets, and even 'exclusive' hair combs and chopsticks. They are fully aware of the excessive relative cost of the purchases, this being part of the appeal. Elsewhere, newly affluent Chinese mainlanders in Hong Kong queue to spend US$500 on a shirt or US$1,000 on a handbag.

Such behaviour intrigues me; not necessarily in the spending alone (which is neither for me to advise nor to judge) but in its attitude towards those still in need. My naïve assumption was that someone who had raised him or herself out of a very low standard of living would empathise more with those still suffering; but this is often not the case. They may have both personal insight and connections to those who remain in hardship, yet an impersonal detachment regarding helping them. My Chinese friend has little sympathy for the poor in mainland China. This attitude seems to be common in all the places I have worked and visited.

If this is sounding familiar, consider what we learned earlier, where we looked at the relationship between people and their energy supply – particularly with grid electricity. There, the realities of consumption, wastage, expense and pollution were hidden from the immediate consciences of individuals. The interface is no more than a plug here, a button there; the payment under a pile of bills or out of sight in a direct, regular money transfer; the motivation to question or improve matters satiated and suppressed.

Here also we see both individuals and groups displaying an elaborate denial of their wealth in context. Even direct experience of poverty does not equate with empathy with the poor, and empathy alone does not translate into motivation, let alone active engagement or change. People have an uncanny ability to separate themselves from the meaning of their own situation in relation to others, and dissemble on matters of wealth or spending. Disingenuous, perhaps; distasteful, possibly; but all too human; and as such, this may help explain, in conjunction with more familiar political and economic factors, the imbalances and injustices of our world.

And wealth is addictive. As soon as we taste it we want more; and a disconnection grows between 'us' and 'them'. This is not to ignore that some of the wealthy are fantastically generous to others; thousands of philanthropists make a truly positive impact. Unfortunately, real cultural change requires tens of millions of people, motivated in unison, to correct global inequality.

I have seen examples of this psychological divide across the globe. Kigali, the capital of Rwanda, where, even 10 years ago, the only expensive vehicles on the roads were owned by foreigners, NGOs or were government registered, now has Porsche Cayennes and S-Class Mercedes on the streets. A friend of mine worked for a Rwandan with a $100,000 Range Rover complete with white leather seats – perfect in dusty Rwanda! Some of these rich – there is no other word for them – live either in the suburbs where they grew up, or within an hour's drive. Yet there are children dying of malnutrition almost literally on their doorsteps. Visiting one such rural village, the wealthy family's car (there seemed to be just one nice house) would have covered the cost for all village families to have concrete housing and water access.

The upwardly mobile behaviour of people leaving poverty behind represents who we are as a species. Many look at this localised contrast in wealth with condescending eyes. But in reality, their only difference with most of us in the industrialised world is that their neglect of their neighbours is more literal and more obvious. When 2 billion people in the world earn less per day than the cost of a cup of coffee or a pint of beer, none of us really has any excuse; the wages of aspiration are universal. Similarly, though, it is important to recognise that the richest in the world also make a statement about us as a global society. They also influence wider attitudes to wealth accumulation, with money as its principal unit of measure.

> The upwardly mobile behaviour of people leaving
> poverty behind represents who we are as a species

Value and the power of money

The prevailing mantra of our times is consumerism – a culture driven by marketing that often depicts the dream spoils of wealth. It brings with it a problem which defines us all. Our attitudes to personal wealth and money determine nearly all aspects of our world. As I will discuss shortly, it is this collective attitude, and in turn the specific actions of people and groups, that creates, propagates and prolongs energy poverty.

Of course, money is a necessity in addressing many of the problems of the world. But what counts is how meaningfully and diligently we apply it. I described this earlier when defining the principles of 'Personalised Energy'. If we simply donate money, goods or products, they will effectively carry less value. Whatever the semantics, recipients will not truly value them, at least not in the lasting sense required. Those who can benefit the most from energy services need to be motivated to use them and engage with their worth.

In the coming chapters we will see promising examples of progress in the developing world, in which those of very limited means are given a chance to access finance. Such schemes – helping poor people to help themselves – show a spirit of innovation with which the energy sector would do well to keep pace.

For this is the dynamic that informs what follows. In the rich world, people take for granted that they are spoon-fed bountiful energy services drawn from central grids. They may only be shaken from their lethargy once that energy supply comes under threat and the lights go out. And loss *is* a motivator, no doubt. But a greater incentive by far comes with the ability to *pursue and realise financial gain*. For, as we have said, this brings more than money or energy. It is the promise of a better life. It is the promise of inclusion, a dependency-free stake in an emerging, economy; where locally empowered individuals, households, villages and districts stand taller as their own energy providers. Taking care of business.

Endnotes

1. https://www.weforum.org/agenda/2015/05/why-are-most-of-the-worlds-hungry-people-farmers/   and   https://nextbillion.net/why-the-worlds-smallholder-farmers-are-going-hungry/
2. Acemoglu and Robinson (2013). This book has much to say on these underlying movements. Coming from the field of energy, I feel particular affinity with its long-term outlook; I also quote from Jared Diamond's 2012 review of it in the *New York Review of Books*.

CHAPTER 15
Finance, technology and people

Abstract

Money for development is an emotive subject, not least because so much of the stuff is already being spent.

This chapter is a basic overview, which I hope will not only demonstrate the mechanisms to make capital available, but will also provide some insight into why they do not necessarily deliver the expected results.

The idea, as with my overview of poverty in the last chapter, is to set the scene for more insightful discussions on technology, specifically autonomous energy products in the developing world.

It is important to understand that energy poverty is not defined exclusively by money or technology; and why it will persist if we continue with the prevailing approaches to 'alleviating' inequality.

On one hand, organisations are going to great efforts to develop and provide finance for the poor; but on the other, they are spending that precious finance on low quality products and services.

The barrier to alleviating energy poverty is how we choose to apply our money, our efforts and the best of the available technology.

Selling Daylight – that is, personalised energy services – is a commercial strategy. Its philosophy is, first and foremost, to bring high quality and versatile energy services into energy-poor regions, through commercial sales, to those with money. These services will then be more accessible to everyone. But a frustrating reality for millions of people is lack of capital, for purchasing items which could subsequently generate further funds. At one end of the global wealth spectrum, there are a few hundred people worth billions of dollars each. At the other end are billions of people, each struggling to raise tens of dollars. Those tens of dollars are the critical ones, because most items of quality and value – foods, medicines, educational aids, communications and lighting – require capital expenditure of some kind.

Access to finance

Development, as with any effort to combat poverty, critically depends on access to finance. This is widely documented and understood; my own experience has confirmed the volume of energy service products – PV among them – which are purchased as a result of some kind of financial assistance. Of

http://dx.doi.org/10.3362/9781788530705.015

course, simply making better products will make no difference to those who cannot afford them in the first place. I shall come to this later, and describe the mechanisms which can negate the need for capital expenditure by the user altogether. Initially, though, let us look at making quality energy services as accessible as possible.

Capital funding mechanisms

The purchase cost of an electrical energy product or solution (that is, the capital cost required to take ownership) can be the result of savings, or of external financial mechanisms. Providing capital to a buyer can take many forms. It can be a grant without the need for repayment; a loan that is repaid over a defined period of time (with or without interest); or as part of a tariff system, whereby the user receives the product at a fraction of its usual purchase cost, then subsequently pays instalments over a period of months. Alternatively, the energy product or solution can simply be rented, without the user ever actually coming to own it.

Money for development is an emotive subject, not
least because so much of the stuff is already being spent

Access to finance for poor people is complicated. The availability of finance facilities (formal or informal) is not the only issue; people's attitudes, characters and customs play a significant part. It may depend on whether certain communities actually use money; or if they do, their ability or willingness to enter into a savings or repayment plan.

Savings, trust in institutions, and stability

It is human nature to appreciate something more if effort has been involved in acquiring it. This holds true for all of us who have ever saved up for a purchase, especially over an extended period of time, or at the notable sacrifice over other benefits. For this reason, it is preferable that an energy service solution is purchased with the user's own savings, given our vision of 'Personalised Energy' based on personal value and engagement. The practical limitation on savings is that only around 40% of the adult population in the developing world have a bank account (Demirguc-Kunt and Klapper, 2012). Mobile phone based banking is gradually changing this. But there are bigger challenges, not least in the limited use of actual money; not to mention a deeply ingrained and culturally entrenched distrust of institutions.

A village community I worked with in Papua New Guinea do not use money for anything other than small transactions with outsiders. A friend and I had a conversation with a pastor in Uganda regarding savings and bank accounts. His initial reaction was to laugh. He explained that most people don't trust banks; either the organisations themselves, or the government, would steal the money (historically not uncommon across the African continent). He also explained that people do not expect to live to old age, and they don't have 'spare' money to put aside. Children are effectively their savings plan.

These cultural values carry a measure of fatalism, because they are played out against a world full of critical and potentially devastating external risk. Exchange rate variations can wipe savings away in an instant, as can the devaluation of currency against an increase in imported goods. At the same time, the trade practices of dominant nations with their poorer 'partners' only serve to keep people in poverty, and at the bottom of the development ladder. Even when people do have some financial capability, overwhelmingly it is the initial amount of money they hand over that defines, for them, the validity of their purchase. Neither ownership cost, quality of service nor operational life will be uppermost in purchasers' minds.

This harks back to what we saw regarding value, as opposed to cost. Tens of billions of dollars are still spent on kerosene lamps, despite 30 years of solar lighting. This is due in no small part to the low capital cost of the lantern (some are home made out of food tins). As we have noted, this has led to an attempt to make modern lighting products cost as little as possible, at the expense of quality.

People do not expect to live to old age, and they don't have 'spare' money to put aside. Children are effectively their savings plan

---

Grants and subsidies

To enable access to modern energy services, a range of formalised purchasing mechanisms exist for addressing the barrier of capital cost. In theory this should establish a degree of quality control into the process. What is the reality?

Grants can cover the cost of the energy product; either in total, or in part as subsidies. In principle, either is a good way to purchase PV, because much of the ownership cost is upfront. However, grants and subsidies are often short-lived and subject to political objectives. The quality of the service and training of recipients have a secondary importance, behind the primary, head-line-friendly outcome; meeting a target and ticking a box to demonstrate a given percentage of the population with 'access to modern energy services'. More importantly, simply providing a ready-made solution will fail to instil a sense of ownership with the user or group. They are therefore far less likely to be efficient with its use, to care for the product or even operate it correctly.

Grants and subsidies also undermine the ability of the private sector to provide such services on the free market, limiting consumer choice and preventing the development of local industry. The result is a collapse of the local supply and support market after these grant programmes have inevitably run their course. The most immediate effect of this is a breakdown in addressing even simple maintenance and support issues; and this will reduce the operational life of the energy service.

Every time I visit a developing world community, I see PV products discarded due to very simple issues – dirty panels, dirty plugs or sticking switches. Each of these can be resolved in less than 5 minutes with very basic knowledge. The best subsidies take account of this wider, long-term cost; they are carefully targeted to help overcome specific challenges, such as subsidised training for operation and maintenance, or marketing assistance to promote quality and value propositions. However, after all the hard work that goes into accessing finance, that money is often wasted on inadequate goods: the most prominent example being solar lights.

Money devalues technology

Grant and subsidy programmes often limit the recipient to a list of pre-qualified energy products. Far too often the financiers fall into the same trap as the end users; they prioritise product cost above fitness for purpose and ownership value.

Money (when we interpret it as mere *cost*) is the biggest obstacle to quality and value. Money defines and distorts people's perceptions of value, in both directions: a 'cheap' product can be celebrated and promoted as a 'deal' or a 'bargain', while a designer label handbag is considered high quality partly due to the price tag enabled by its luxury brand.

There are practical limits to this, whichever way you want to stretch it. Once a handbag is a couple of hundred dollars – assuming quality is the design priority – there is not much quality that can be added with an extra thousand dollars. If people wish to spend that amount for what is effectively the market rate of the logo, so be it.

At the other end of the scale we find that quality, rather than brand identity, becomes critical. As cost is reduced, be it for the sake of the sale price or otherwise, the quality suffers. In an ideal world, there should be a minimum quality line, beyond which goods for the poor (or anyone for that matter) should not cross. Of course, there is no such threshold; evidenced by the cheapest end of the consumer goods scale, where things fall apart when you look at them harshly.

So again, we see that the principles of both 'quality' and 'value' refuse to lie down. Money is so incredibly valuable to those needing assistance, yet circumstances, misadventures and attitudes can all conspire to waste it. The trick is to encourage consumers, vendors, manufacturers, financiers and investors to recognise this broader economy of value and to act rationally upon it.

Energy poverty and people

Energy is defined by money. 'Energy poverty', on the face of it a somewhat nebulous term, is defined by people – in particular by our attitudes to money and how we convert it to apply energy technology. What emerges from this conversion is – that word again – a service; the tangible benefit, whose effectiveness depends on how it is used.

How do energy services transform lives? The role they play in social and economic development is so far-reaching that it is hard to know where to start. Lighting, on its own, opens up giant opportunities for education, healthcare, community safety and a world of activities which support the generation of income. Children spend hours each day fetching wood or other fuels for cooking and lighting (as well as collecting water), keeping them out of school. A person cannot read after sunset without lighting of some kind, so self-motivated education is not possible, let alone formal schooling. And this does not only apply to children; it means health education, land management, family planning, hygiene, vaccines (which require cooling) and any form of communication other than face-to-face. Some countries are so hot that it is not practical to try to concentrate during parts of the day, so people prefer to study while it is coolest, in the middle of the night. But this is only possible with artificial light. Mobile phones, internet access, banking, radio and TV are all dependent on energy. The list goes on.

'Energy poverty' is defined by people – in particular by our attitudes to money and how we convert it to apply energy technology

---

I take the simple view that, like it or not, our modern world runs on electricity. We could eat and drink, maybe, if we did not have electricity, but nothing we do today in business or commerce can happen without it. Our standard of living is certainly built on electrical infrastructure. Modern communications; powered transport (in which we include fuel logistics); recorded, broadcast and much live entertainment; and digital technology (fixed or portable computing, the internet). All are dependent on electricity. If we fail to extend these services to those who do not have them, we are denying them access to our world. And our world is the one with all the opportunities, all the money and what we consider to be the basics of life: education, healthcare, business tools, safety and security. It is bewildering to consider that those last two sentences could have been written in a science fiction book and be referring to two different planets. But the people we are talking about are most likely within a short plane journey of most of us. It is becoming clear that there are selfish reasons to address energy poverty, and given the human condition, these are probably the only viable motivators. And the most powerful of all selfish motivators is profit.

International recognition

Energy's critical role in development has been recognised by the international community. For the avoidance of doubt, these examples make the link between energy and poverty explicit, and subject to strategy and policy:

- 'Expanding access to affordable, clean energy is critical for realizing the MDGs (Millennium Development Goals) and enabling sustainable development across the globe' (Secretary-General Ban Ki-moon; United Nations Development Programme, 2011).

- The UN General Assembly Member States have unanimously declared 2014–2024 as the Decade of Sustainable Energy for All. The Sustainable Energy for All Initiative has three objectives by 2030: ensuring universal access to modern energy services; doubling the global rate of improvement in energy efficiency; doubling the share of renewable energy in the global energy mix.
- None of the MDGs can be delivered without access to modern energy services for the 1.5 billion people who today live without it (World Energy Council, 2011).
- About 110 million households and more than 10 million microenterprises across Africa have no access to electricity which greatly curtails their socio-economic activities once darkness sets in (International Finance Corporation, 2012).

There is no meaningful disagreement about providing energy services to the poorest and most vulnerable. The differences concern how best to provide those services. I shall come to explain, in purely logistical terms, why I consider commercial manufacturers essential for supplying high quality energy services to energy-poor regions. For now, it is important to understand that energy poverty is not defined exclusively by money or technology; and why it will persist if we continue with the prevailing approaches to 'alleviating' inequality.

Energy-poor regions are not defined exclusively by money or technology

The attributes of energy poverty

How did we get here? Rather, what is the combination of financial, technological and human factors that ends up creating a situation where vast numbers of people go without the benefits of energy? The many financial mechanisms for purchasing electrical energy services *should* be making a significant impact on energy poverty; particularly in offsetting the expense of fossil fuels used for lighting. Unfortunately, while there is definite progress, we as people impose limitations on how the hard-earned capital is spent; ultimately, this undermines what these energy services are capable of providing.

Electrical services, meanwhile, form the backbone of the modern world. In the past, it was the capabilities of rural people to access finance that often directly defined their subsequent ability to earn a living. Finance allowed the purchase of seeds, fertilisers or the start-up for, say, a locally based microbusiness. However, the present day is increasingly dependent on electricity, specifically for communications and digital interactivity for businesses and social opportunities. Without energy services, one's ability to develop – socially and economically – is increasingly hampered.

Finance, therefore, is now part of a two-stage process, in which energy services are required *before* a person's ability to earn a living increases. The quality and value of the energy product therefore has a significant bearing on the finance capabilities of the buyer. This is definitely the case for PV-enabled electrical services. (Here, I am not referring to those in acute poverty, dependent on immediate assistance; but to those with very small amounts of money – above US$1 per day, but maybe not much above – who struggle to make ongoing progress, and who are held back from economic development.) How people spend money on energy, with or without third-party assistance, is largely determined by the quality of available products and service solutions. On one hand, organisations are going to great efforts to develop and provide finance for the poor; but on the other, they are spending that precious finance on low quality products and services.

What about the technology itself? Increasingly, the poverty debate calls for new technologies to solve the problems. Technology for learning, healthcare, communication and business have definitely been restrictive to developing nations in the past. But that is no longer the case. Just as the persistence of poverty is not due to a lack of available money, energy poverty is not due to any technological constraint. Rather, today it is the *application* of technology that contributes to widespread energy poverty – and as such, poverty in general. The developing world is a graveyard of archaic and inappropriate technology, littered with fax machines, dot-matrix printers, ungainly CRT monitors and photocopiers the size of a fridge (with the power consumption to match). Computers run 10-year-old, unsupported operating systems and virus-riddled software. Consequently, otherwise simple tasks become day-long challenges in patience.

However, the specific choices and the application of the technologies have very often hindered progress. I learned this first-hand in Rwanda in 2005 as I demonstrated how to use PV to charge laptop computers, then stood embarrassed as the students struggled with the inadequate equipment, faced with the Windows operating system, no other software and no internet connection. We now have intuitive interfaces on tablet PCs and all the educational material anyone could need, as long as a good internet connection is available. But even the best of these devices depends on support of some kind, whether it is online, or served by skills on the ground. Supplying computers into regions without skilled people or online access is highly likely to result in problems.

> Just as the persistence of poverty is not due to a lack of available money, energy poverty is not due to any technological constraint

This may sound a familiar tone. I described earlier how rich nations import fossil fuels dependency and central generation into places where they restrict development. We run a similar risk if we make assumptions regarding technology. Accustomed to skilled support, internet access and (of course) electricity, we don't stop to consider just what it really entails to use these devices. It is too tempting to impose our way of doing things into vastly different environments, only to wonder why the results don't match our expectations.

This is the challenge: to adopt the existing technology we have, alleviating the frustrations of old equipment; but to adopt it with a view to how the tools will be used. The correct technology exists now, in our transport, communications, medical services, education and pretty much every aspect of our comfortable lives. It is ours to command as we choose.

And it is the masses of *users*, rather than the innovators, who are effectively the financial enablers of new technologies. It is the *people* who turn them into valuable tools, who define their application, their function and their value; and in doing so, define the attitudes of the society in which we live. The society we have created happens to permit and promote energy poverty – and will continue to do so.

The value of money and technology; microfinance

There are now widespread facilities to enable loans to poor, rural people in the developing world. Microfinance, known also as microcredit, is a powerful enabler of access to energy services. There are variations in its provision mechanisms, but they share a common factor in that they displace the restrictive attitudes of traditional banking. They are also a growing example of how people undermine the effectiveness of both money and technology.

One problem with conventional banking is that, to reduce risk, banks need to be sure of collateral before providing a loan (the house, for example, when a mortgage is taken out). The world's poorest people have no such collateral – at least, none of interest to the mainstream banks; so they cannot start to invest in the energy services that could help them out of poverty.

Rich nations import fossil fuels dependency and central generation into places where they restrict development. We run a similar risk if we make assumptions regarding technology

---

The general concept of microcredit is that small loans are given to people without any collateral. The small size of the loan means that the bank does not take a large risk, but repayments are expected. Generally, microcredit services should be offered locally in low income areas and appropriate to local conditions – that is, affordable for most of the population.

The most well-known and successful example of this is the Grameen Bank in India. Relying upon mutual trust and accountability, they bring financial services within the reach of the poorest people, providing modest loans towards services which help increase income (thereby facilitating repayment). One such scheme is the 'pay-for-use' mobile phone. A participant, or 'entrepreneur' invests in a phone using a microcredit loan. Phone calls are then sold to other people (the customers) in the local area to raise income to pay back the loan, as well as generating an income for the phone owner. The Grameen Bank mainly supplies microcredit to women (97% of the borrowers are women) and they have a near-total loan recovery rate (Grameen Bank, 2015).

Many such initiatives are in operation across the world. Some are more locally tailored – and therefore less replicable – than others. For example, where Indian citizens do actually have some collateral in the form of gold jewellery, this can be used to gain higher value loans. One particularly interesting and widely employed format is the 'group microfinance' model, which illustrates the community nature of poor regions, a characteristic which is vanishing in so many locations in the industrialised world. The first member of a small group (four to eight people; again, usually women) is given a loan, but the next group member does not receive their loan until the first recipient has repaid all or a significant proportion, of their individual debt. The third group member is dependent on repayment by the second, and so on. In short, peer pressure is used as the formal collateral, as a single person is responsible for the group's credit (even though it has been reported that the money is often shared within the group throughout the process).

But once again, despite the proven capabilities of microfinance programmes, without a doubt the most widely practised method for addressing the barrier of finance for energy services is simply to reduce the purchase price. As we have seen, because suppliers and purchasers of PV products are so driven by cost, this has to a large extent defined the quality of products supplied to microfinance organisations. This has only served to extend the cycle of spending – by those very people whose money is so precious.

PV is a high capital, low running cost solution. Ultimately, it is a good fit for microfinance: all, or at least a very high percentage of the cost is upfront; and it can deliver very high reliability if designed and supplied appropriately. By contrast, kerosene, diesel and battery charging are the exact opposite, relative to what they provide. But in spite of the lessons learned and all the effort of enabling access to capital finance, the unfortunate reality is that so little care is taken on the quality and the subsequent value of the products it is spent on. Bluntly, PV products aimed at microfinance have not been sufficiently reliable to date. So owners have ended up losing their service – lighting, TV, radio or phone charging – yet still paying for the system. Payment defaults are generally low until the product fails.

We have seen how two great theoretical solutions – microfinance and stand-alone PV – should represent a genuine life-changing combination. Yet they are hampered by the very people making the effort to help. Overwhelmingly, this is a reflection of the monetary focus we all share; a misguided reliance on cost as an indicator of quality. With so much opportunity in this sector, this fact alone should stand out. The barrier to alleviating energy poverty is *how* we choose to apply our money, our efforts and the best of the available technology.

CHAPTER 16
Scale, diversity and people

Abstract

The sheer number of people we need to reach and their diversity of circumstances is often cited as a principal, insurmountable barrier to our honourable efforts to solve the problems of poverty and modern energy access. But over the last 25 years we have supplied modern products and services across the globe to numbers well in excess of 2 billion people.

For me it doesn't matter if you believe our focus is diverted by global catastrophe, by politics, corruption, religion, financial restraints or simply because we are human. We must face the fact that, to date, our collective efforts to address injustice are not working.

This chapter focuses on the need to stop pretending that our big global problems are someone else's responsibility. Its purpose is to make it clear that motivating people is at the heart of any solution.

There's one hell of a lot of people in the world.

This obvious fact has long been invoked to demonstrate the reality of enduring global poverty and inequality. Historically (and risking gross understatement), there is a troubled legacy in attempting to transform the lives of large populations. Behind the straplines, we must accept not only the sheer numbers of people, but the diversity of their cultures and the variability of their circumstances. If we achieve this, we are entirely capable of reaching, both logistically and emotionally, the people who most need us to do so. As with money and technology, it is a lack of motivation or collective will which holds us back.

The numbers describing poverty

First, a look at scale. The highest and most striking figure is around 3.6 billion. This is the number of fellow humans without modern cooking or heating, access to hygienic water, sanitation or electrical services.

Food and cooking, heat and water; these are essential to life; and in fundamental terms are more of a priority than electrical energy. I know that there are solar thermal (space heating and hot water provision) technologies widely available, as well as solar cooking product solutions. Some professionals speak well of solar cooking, others less so. Its effectiveness is limited to places with good, strong predictable sunlight. Elsewhere, many organisations are trying

http://dx.doi.org/10.3362/9781788530705.016

to address the issues of heating and cooking using cook-stove technologies, biofuels and fuels from waste. But this is not my primary focus, and I will defer to the professionals in those fields.

Similarly, as an electrical specialist, I know that PV holds great potential for irrigation, water access from underground and for filtration. Electricity as such does not address the immediate issues, but I am not proposing to add specific value to the organisations in that field. What, then, do solar-enabled electrical services bring to the solution? In other words, what does my proposal have to do with the war on poverty?

Instead, I wish to focus on electrical services *for those who can benefit from them the most*, particularly in the areas of education and health. Getting this right – establishing a reliable and predictable local electrical infrastructure – is an essential building block, which will massively benefit the essentials of food, cooking, heating and water. The disclaimer, as always, is that electrical services must do their job properly through conscientious design, deployment and support, or they will only compound the fundamental challenges.

Two numbers are habitually quoted to illustrate electrical 'energy access'. The first is 1.3 billion, and refers to people without access to electrical services. This is generally assumed to mean grid electricity, although strictly we must acknowledge alternatives, such as local networks powered by diesel generators and renewable technologies. Secondly, a figure of 2 billion refers to those requiring *dependable* electrical services, where the utility grid is an impractical distance away or cannot be relied upon for businesses, education, healthcare provisions and other key services.

The numbers are necessarily broad in scope, but as conveniently quotable figures they distract from their own bewildering magnitude – so familiar and so large that they almost certainly possess no meaningful impact anymore. Perhaps they never have done; but in short, a good chunk of the world's population – fellow humans – do not have basic or reliable electrical services.

3.6 billion is the number of fellow humans without modern cooking or heating, access to hygienic water, sanitation or electrical services

---

Stadiums full of people

We can continue a while with this abstraction of scale. To recap: 3.6 billion people cook and get light and heat from burning biomass – wood, dung and crop residue; 2.6 billion people do not have access to basic sanitation; 1 billion people lack access to safe drinking water. (In 2010 half of the world's hospitalisations were due to contaminated water – infections, toxicity and radiation.)[1] And 900 million live in slums. Are these numbers meaningful? We are unable to quantify, much less qualify, the numbers of people who need help, and this is a major barrier to envisaging or communicating solutions. I, like many, have struggled to even relate to the scale.

Think of how many people live in your street, how many attend your school or work in your building. What is the population of your town, your city, your county, your state or your country? And how do they compare with these gigantic figures mentioned above? Try the following simple exercise in picturing as many people as possible. Be mindful of the point at which the quantity becomes too abstract.

Take a large sporting event and try to picture a stadium full of people – the Super Bowl, FIFA World Cup final or the Olympics opening ceremony. Imagine an aerial view of one of these events. Let us assume, for the sake of easy maths, that the stadium holds 100,000 people. These are among the largest gatherings of people that most of us can visualise in an enclosed area.

Now try to replicate an identical stadium immediately to the right of the original. Add four extra stadiums above each, so we have two columns of five stadiums neatly lined up next to each other. This is what 1 million people look like. Double these to picture 20 stadiums in all; then double it *again*, and we have reached the 4 million mark. This probably nudges a reasonable threshold of clarity for most people. Nevertheless, we will soldier on, and try to double the number of stadiums now *in each column* to 10 instead of 5. This gives us an imposing mental image; eight columns representing 1 million people each, lined up next to each other to make 8 million people in 80 stadiums.

Have the numbers become somewhat abstract yet? It is tempting at this point to use something like nuts, coins or beads to represent the stadiums. Continuing, try to picture 10 times that amount – 800 stadiums. We now have 80 million people. Repeat this, giving us 800 million people in 8,000 stadiums.

This might start getting tricky, but stay with me. As a final step (I promise), take this quantity and add the same amount again – another 8,000 stadiums, *then the same again* – another 8,000 stadiums of people. We have reached a blurry and confusing representation of the number of people on our planet that suffer due to a lack of basic human needs such as clean water, nutrition, healthcare and freedom.

Humans are not capable of picturing this quantity of people with any kind of realistic perspective. Using this analogy of a sporting arena as a unit of mea- sure, at which point did the numbers become meaningless? Four stadiums? 10 stadiums? 100? When did you turn the page looking for a handy infographic with stylised ranks of stadium icons (a shortcut I have deliberately omitted)? And last but not least, when did we forget we were counting people subsisting below a reasonably acceptable quality of life?

Making sense of large numbers

So how can we make sense of trying to feed this number of people, or provide them with clean water? (Here I will abandon the sports stadium analogy, with its captive market for insultingly overpriced fizzy drinks and junk food.) The blunt answer is that we cannot; the sheer scale of the problem is too massive and we have no ability to relate to it. So many well-intentioned people try, particularly when they are idealistic and young, believing that this poses no obstacle. But they struggle as the complexities of scale, distributed geography and culture begin to dawn. In time, apathy takes over and people simply dismiss the issue from their everyday lives.

And this is understandable. Run the numbers through your mind again: nearly a billion people in poverty; 1.3 billion people without electricity; 2 billion without access to reliable energy services; 2.6 billion without sanitation; and a global population of 7 billion, predicted to rise to 9 billion by 2050, largely distributed in energy-poor regions. And all of this against a background of fossil fuel depletion and dependency, and the ticking clock of climate change.

No single company or sector can address the challenges in isolation. To date, governments have not proved capable of combining efforts for the future of the Earth and environment. The collective will to save a bunch of poor people seems far-fetched at the time of writing. A new international agency is likely to move too slowly, while the established ones have proven ineffective. Even if a well-positioned body such as the World Bank were suddenly to adopt an inspired strategy for propagating energy services to poor areas, it no longer has the credibility among the thousands of stakeholders essential for implementation and long-term management.

The truth is, a problem of scale requires a solution of scale. If we wish to address a problem facing billions of people, we need tens of millions on the job, active in the far corners of the globe where communities are facing hardship, in every essential service sector and in every stage of the supply chain. What is more, they must be motivated continually for decades, not just momentarily energised by a short-term campaign. Likewise, it is vital to engage those organisations with proven capability of meeting such challenges in the field. But once in these far-flung areas, we find that with scale comes another set of challenges.

The meaning of diversity

I will reiterate here: the problems of poverty and modern energy access are not due to the number of people we need to serve. Over the last 25 years we have supplied modern products and services to numbers well in excess of 2 billion people: personal computers, televisions, media players, gaming consoles, washing machines, refrigerators, freezers, air-conditioners, ovens, automobiles – along with their associated spare parts and support and service infrastructures. Instead, the barrier to propagating services and products, including energy services, is *diversity*; the varying geographical and cultural contexts in which populations live and work.

> The problems of poverty and modern energy access are
> not due to the number of people we need to serve

Diversity of location

Few people can adequately comprehend the sheer divergence of environments in which people live, or the range of challenges posed purely by terrain and natural phenomena. As geographical and logistical challenges differ from one region to the next, so too will the relative effectiveness of environmental projects. Put bluntly, it is too easy to assume that a technical solution will work universally on account of its ingenuity alone; it is equally easy from a comfortable distance to ascribe a type of climate or landscape to entire regions, even continents. A clean water project that works successfully in one region might be based on extracting water from deep underground, whereas some communities have an abundance of water. Here, our common assumption of a dry landmass for the majority of Africa is a pitfall to avoid. Besides this, it's what *lives* in the water that is the problem.

Diversity of culture

Here, the differences between cultures comes into play, which is why an effective method of communication with one community is not guaranteed to work with another. The priorities and perceptions held by people in different regions may prove as much of an obstacle as the physical terrain in which they live. For example, a clean water project might not work because people do not realise that they even need *clean* water. (The story of the 1854 cholera outbreak in Broad Street in London serves to show how public health can be transformed overnight; by tracing the outbreak to a single source, the solution became a matter of educating the public as much as improving the cleanliness of the environment.)

This is why the efforts of tens of thousands of charities (sometimes hundreds committed to single sectors such as water, education, gender rights or child safety) still fall short of addressing the basic problems of the developing world. It is not the efforts in themselves which are lacking. It is the number of people who need to be part of the same strategy, positioned on the ground in the localities where action is needed.

Overcoming scale and diversity

The sheer number of people we need to reach and their diversity of circumstances are indeed intimidating, and this is often cited as a principal, insurmountable barrier to our honourable efforts. Yes, it is difficult enough to appreciate matters of scale – populations, distances and so on – in their own right. But the more one understands the high *variation* of needs within that scale, the more one acknowledges the localised, nuanced problems in their context, for what they are.

We are entirely capable of embracing the diversity of people in need. We routinely use video cameras and computers in the deepest oceans, smartphones and cameras on top of the highest mountains, digital communications in orbit and beyond. There are no logistical reasons *not* to reach poor people here on Earth, most of whom are no more than a matter of hours away from wealthy populations.

The people problem

My descriptions of our monetary, technological and logistical capabilities have been fairly basic, but hopefully I have expressed what we should all know to be true. Before we can enter into a rational discussion about the solutions, we should come to understand the reality of *why* such appalling inequality and destitution is permitted in our world.

Potential versus reality

Let us fantasise, from the realm of science fiction, that we own a space-age teleporter. We load it with items in one location, press a button, and the items magically disappear, only to re-emerge in the destination of our choice within just a few seconds.

Say we could load it with a 100 kg container of clean water, along with fruit, vegetables, rice, maize, nuts (continuing with all the vitamins and minerals required for good human health), and almost instantaneously deliver those items into the centre of poor and malnourished populations around the world. My question is this. With such technology, could we – would we – end extreme poverty in the world? I submit that we would not. In the vast and epic human story, it is not the momentum of technology, but humanity – we ourselves – who have sidestepped our own responsibility in the name of material, technological 'progress'; the authors of our own problems and the barrier to a better world.

A dim view of humanity? Consider the state of transport in the 15th and 16th centuries when global exploration by sea became established. The wealthy and technically advanced of the time (the two being mutually supportive) set out to colonise the world. When they happened upon peoples with a poorer standard of living, did they set up a supply chain of produce or share technological know-how? No. They stole the produce of these new lands, then proceeded to steal the people.

> There are no logistical reasons not to reach poor people
> here on Earth, most of whom are no more than a
> matter of hours away from wealthy populations

If we were somehow able to explain modern international trade to those 16th century explorers in their time, they would not have been able to comprehend the technological advances that had taken place: commercial fleets crossing from London to West Africa in days instead of weeks, in vastly improved conditions, carrying not a few tons, but hundreds of thousands of tons of goods at a time. Imagine describing the occult wonders of air travel, and the ability to reach poor countries from rich ones in a matter of hours. But would those early maritime explorers have believed that inequality, injustice and poverty would exist on such a scale in the 21st century? Yes! Because during their time it was an accepted fact that the rich controlled the poor. (Diversity between cultures exists in time as well as space; the past, to borrow clumsily from literature, being a foreign country.) Technology was power, and those in possession of it were considered, by themselves and others, superior and deserving of power and control over other humans. International high-capacity shipping made the poverty gap worse. So, irrespective of its image of popularising tourism, did air travel.

Turning to modern communications in general, the telegraph, international mail, intercontinental telephony, fax machines, email, internet voice and video – all of these could have underwritten the eradication of extreme suffering, but this has failed to materialise.

In fairness, there are exceptions. Healthcare, literacy, education, clean water, sanitation and nutrition have improved the living standard of millions of people; either by external organisations or by the needy themselves, using modern technologies in ingenious ways. But after hundreds of years of international shipping and many decades with international air travel, the inequality in our world remains disjointed. We can deliver goods across oceans and continents, in less time than it takes to struggle the last few dozen miles, due to poor local infrastructure.

Television has shown us images of the suffering poor for at least four decades; radio has described them for even longer. For at least the last decade, the internet has allowed us to verify the disparity between rich and poor for ourselves, along with the damage we are causing to the planet, forests, species and those humans supplying our goods. We can enjoy this on giant plasma displays, in high definition and 3D if we so desire. Those serving us are no longer a different skin colour and building the infrastructure in our countries. They are now in factories or rural settings in a multitude of global locations, slaving at making goods for our entertainment, our comfort and our consumption.

The world is not how it should be

For many years of my career, since I began trying to fathom a way to deliver high quality, cost-effective solutions to energy-poor regions, I used to face an invisible, yet seemingly immovable obstacle. I insisted on thinking of the challenge only in logical terms, and of human behaviour as moral, or at least rational. This was a mistake.

People with power define both the relative value of money and the effectiveness of technology. But, when you consider the wholesale depletion of the Earth's resources, social inequality, the suffering of the poor or environmental collapse, one must conclude that there is *no logic to how we act*. I did not realise it at the time, but I was looking for a method that fitted how the world *should be*, rather than one that fits with how the world actually is. Myths of solidarity, spurious notions of the wisdom of crowds, and phrases such as 'global village' exist to make us feel good about ourselves. But we are not logical as a global society any more than we are as individuals, and our world is not how it should be. Nor, probably, will it ever be so.

The resources riddle

It is a continuing irony that we spend most of our money on items which are sourced from the developing world. Commodities such as coffee, tea, cotton, maize and rice are produced at globally significant scale by poor countries. More striking are the valuable gold, silver and gem stones which have been prized so highly throughout history. (Those 16th century colonists did pretty well in this regard.) Rich societies in the modern age are fuelled by oil, coal and uranium, sourced in no small proportions from poor countries. Just to add insult to injury, new energy and transport technologies, smartphones, computer and communications infrastructure – the very devices which are accelerating the gap between the haves and have-nots – are dependent on rare earth elements, and there are no prizes for guessing where most of them are located.

Indeed, resource-wise, the world is back-to-front. If you were to look at the world's nations purely from a resource perspective (what can be grown in, or mined from, the ground, and the efforts of working people *on* the ground), and the natural energy resources available (sun, wind, ocean, geothermal), the wealthiest country in the world would be the Democratic Republic of the Congo, followed by a number of South American, Asian and African nations in the top ranks. This is a contradiction of fortune, as these nations are most commonly associated with lists measuring degrees of suffering.

Illogical and irrational

On a personal level, we are prey to irrational interpretations of the world, and cognitive biases in the face of facts. The rise of obesity, for instance, is regularly attributed to food and drink companies, to governments or to a general lack of information. But it also concerns knowledge, personal responsibility, moderation and self-motivation. (Managing my own body weight has been a lifelong endeavour, so I understand the challenges of personal control in the face of temptation.) Smoking, meanwhile, is more straightforward, the evidence of its effects being more prominent and proven over time. Even the most graphic of health warnings on cigarette packaging will fail to stop the most committed smokers. Fully aware of the cost, the risks and the ways to quit, the problem they face is not one of information, awareness or the physical ability to change their ways. As with obesity, there is an element of personal sacrifice, yes; but also a need for continued personal development.

Can this be true also of other behaviours? Having worked with or around very wealthy people, I understand some of the temptations for material gain, but also for signs of wealth and status. Is it possible that the terms 'retail therapy' and 'shopaholic' are similarly ways of suppressing personal responsibility? We live in a world with around 1 billion obese citizens; and, conversely, a similar number classified as in extreme poverty. In this world, people are more offended by a hand gesture or using the wrong fork at a meal than by the facts of extreme inequality and suffering. What does this say about modern civilisation? Here, perhaps, we find an inkling of a cause. Where there is room for personal investment, personal responsibility or self-actualisation – something implying action or change – the human mind will find a way to bypass it.

There may be many reasons for this. Much is reported of 'donor apathy' or of how our compassion reduces with the scale of the problem; why our altruism and empathy are higher for one child than for two, and how it abates when considering those outside the family or tribal unit, especially when in their thousands (evoked by the chilling phrase 'genocide neglect'). But fundamentally, where large numbers of people exist in an affluent society, the mechanisms of conformity, and the cultures by which they create, acquire or maintain status, demand that certain rational 'fallacies' or institutionalised biases to come into play. In other words, *Homo sapiens* has become pretty skilled at changing the subject.

Make up your own mind on this. For me it doesn't matter if you believe our focus is diverted by global catastrophe, by politics, corruption, religion, financial restraints or simply because we are human. We must face the fact that, to date, our collective efforts to address injustice are not working.

Motivating millions of people

The focus of this work is energy; specifically, a solution for eradicating energy poverty. How does it relate to the wider discussion of poverty in this section? My specific convictions go something like this. Accepting, before all else, that we – humans – are the cause of extreme global poverty and inequality, we can rule out a couple of common excuses from the equation. A trillion dollars spent in Africa to date on aid and development funds reveals that we do not have a shortage of money in the world. While it seems clear that the *application* of money is seriously in question, nobody can claim that the funds cannot be raised, in any sector, by any means.

Next, I know for sure that technology does not constrain us. We take expeditions to the poles of the planet, into space and deep into the oceans. There is no technical barrier to supplying autonomous electrical services, especially in very sunny locations to people who require relatively small amounts of power. And there is no physical reason impeding the progress of efficient, durable and reliable products and services, given the correct design, procurement and deployment. The knowledge and the experience are both to hand.

The ultimate obstacle to progress in this field lies in human behaviour; specifically, the need to stop pretending that our big global problems are someone else's responsibility. It is clear that motivating people is at the heart of any solution. Eradicating energy poverty requires that considerable numbers of people must become self-motivated for a sufficiently long time with a compelling purpose, if they are to succeed in a common goal of change.

What form does this take? I shall argue that ethical, moral or environmental arguments are not sufficient drivers to motivate people to act. Nor is the potential to save money over the longer term; or even using the rhetoric of defence, say, to reduce the threat of terrorism.

An independently motivated task force, acting internationally, is not feasible without the ingredient of transaction, of self-interest. I propose that the most effective way to motivate people is with personal gain; principally with direct financial profit.

Endnote

1. https://gridarendal-website-live.s3.amazonaws.com/production/documents/ :s_document/208/original/SickWater_screen.pdf?1486721310

CHAPTER 17
Motivating millions of people and giant companies

Abstract

Poverty can only be alleviated and development sustained with regular, predictable and continued effort. The only way to achieve this level of commitment from sufficiently high numbers of people for sufficiently long is with the motivation of financial profit.

I suggest that there are two generic candidates for engagement: competent global technology manufacturers; and high volumes of local people, relative to the location of need.

Here we look at some of the fundamental approaches which we must adopt for successful energy services to have meaningful global positive value. We then look at the financial mechanisms with which we can motivate companies and individuals alike. This includes the essential task of making environmental responsibility profitable.

This is not just about making products – and especially not about making cheap products. There are also practical trade-offs involved when scale, diversity and 'Personalised Energy' are considered in context.

People are the principal obstacle to eradicating energy poverty. This is due largely to our limited ability to process the facts of poverty; our in-built propensity to bias, an advanced aptitude in denial, and ultimately a retreat from social responsibility. The challenge of supplying autonomous energy services to the developing world is therefore very simple to state: *motivate people*. By this, I mean more than having people sign a petition, 'like' an initiative, even donate money or take part in a protest. What I mean is encouraging people into sustained action; humans of all disciplines and denominations applying effort for a substantial part of their waking lives – in terms of hours in the day, days in the year and years of able life.

'Selfish desire' can ensure sustained effort

First, the hard part. Why should I care? Why should the relatively comfortable people of the world be concerned about poverty – to the extent of doing anything about it? Does it matter? Apparently not. The industrialised nations of the 'global society' appear to be slow at tackling their self-made environmental

http://dx.doi.org/10.3362/9781788530705.017

problems, let alone the inequality of means on a global scale. We have come to accept in recent years that there is an integral relationship between fuel dependency and our security. We give financial support to countries within which our enemies manifest. We are powerless to take real action, for fear of losing those precious commodities which underwrite our standard of living and the operational basis of our societies.

People remain in poverty partly because of their lack of energy, and nations are stuck in poverty because of the massive financial and political burdens of fuel dependency. The situation is arguably worse now than it has ever been. Industrialised nations have poverty, suffering and inequality in their own cities: problems which are well within their material capabilities to address, but outside their capacity to care as a human collective. So, if the ethical or moral plea is ineffective, and we are not sufficiently motivated to protect the immediate habitat of our own children, now or in the near future, what else is there?

What remains, perhaps, is the very thing which has driven social and economic growth in the first place. What motivates people in today's material world is the opportunity for profit; particularly with money, sometimes with power and sometimes with positive notoriety, fame or adoration. An appeal, not so much to the conscience, but to the self, to the ego. And to material gain.

It is no revelation to say that money motivates people more than anything else. What is significant is that we do not like to admit it, either on a personal or an institutional level. 'Selfish desire' may seem a somewhat harsh term, carrying with it a moral taint; but it does summarise a powerful defining element of human society in this day and age.

What motivates people in today's
material world is the possibility of profit

The world is not short of committed or well-intentioned people. They hold a meaningful proportion of global wealth but they make up only a small percentage of the able-bodied population, and are outnumbered by the scale of the problem. Rather, the shortfall is of people who take *sustained* action. Just as charities gain more from predictable and continuous donations than they do from large responses to momentary crises, poverty can only be alleviated and development sustained with regular, predictable and continued effort. The only way to achieve this level of commitment from sufficiently high numbers of people for sufficiently long is with the motive of financial profit.

Commercial organisations, local people

Why should people bother helping the poor? Because they can profit personally from doing so. How do they help the poor? They sell autonomous energy products and services. So, who must we motivate? The challenge involves more than simply telling everyone that money can be made from energy-poor nations. For one thing, it will come as no surprise to those devoid of morals already exploiting these regions.

I suggest that there are two generic candidates for engagement: competent global technology manufacturers; and high volumes of local people, relative to the location of need. Commercial organisations are the most suited to the task. There must be a commercial supply-and-demand relationship between entire nations and the biggest, most capable technology companies in the world. These are the organisations which, more than any government or aid agency, have the ability to fulfil the product briefs at the required scale and speed.

But they will not get very far unless they can hook up with existing, robust, in-country supply and support networks. It is well documented that local people are critical to any successful strategy in the developing world. They should be at the centre of the autonomous energy sector in their region; only they will invest and sustain the necessary passion and concern for what is around them. Once established and self-determined, this locally based ownership and support of the energy service will ensure the most suitable, cost-effective energy products for the people and their region.

In line with our motivational approach, I shall go on to describe why local people are commercially valuable, and how they can be (and will *need* to be) written into commercial supply. The same basic incentive applies: in the long run, people who are appropriately positioned 'on the ground' can enable suppliers to make significantly more money than the standard approaches of broad market assessment, design, sales and restricted logistics.

The profit motive: people and companies

Profit motivates people and companies alike. Although it may seem otherwise, people *do* in fact define commercial companies and shape their destinies. Accepting that the best motivation for people is personal gain, we can recognise that this is the best motivation for companies as well. Taking a look at the commercial world, it is clear in this day and age that any major change on this scale needs to be profit-driven. We will look at some of the financial mechanisms with which we can motivate companies and individuals alike. First, though, there are a few fundamental approaches which we must adopt for successful energy services to have meaningful global positive value. This is not just about making products – and especially not about making *cheap* products. There are also practical trade-offs involved when scale, diversity and 'Personalised Energy' are considered in context.

Scale, diversity and 'Personalised Energy'

The autonomous energy sector requires the most capable and intrepid technology companies of the world to apply their resources. There are practical and logical reasons for this; the sheer scale of need places certain conditions on how we should go about implementing any given energy services. For a start, hardware must be sustainable and environmentally responsible. Solutions must also be 'easy' to distribute. For energy-poor regions, this means utilising existing distribution networks as far as possible. Similarly, the vast diversity of users dictates that our solutions must be versatile, and supported by visual information resources (as discussed in our earlier chapter). Last but not least, any personalised energy solution, if it is to be deployed over giant distances and supported among diverse cultures, has to be robust, and its output dependable.

These criteria need to be met both from a commercial and a sustainability perspective. Each looks to the long term. I want to show that this scale of need, for all its demands, translates into a scale of opportunity. Understanding the above requirements is essential if we are to increase the commercial value for energy services and reduce the costs of supply. We will see that these measures are costly in the short term, but they become increasingly valuable as time moves on.

The autonomous energy sector requires the most capable and intrepid technology companies of the world to apply their resources

---

Sustainability and the environment

There is nothing especially radical in these assertions. The case for environmentally friendly energy services, for example, is decades old. Most of us will be familiar with the rhetoric for leaving a healthy planet for our children, protecting the environment, going green, changing our bad habits and so forth. Aside from the genuine extent of our concerns, much of this rhetoric is co-opted by consumerism, marketed back at us in the guise of lifestyle choices. But none of it appears to make much of a difference, and there is precious little in the way of real implementation to support all the talk.

I joined the renewables industry in 1999 when there was widespread and growing concern about what was happening to the planet. This was becoming more and more a part of everyday media and entering the public consciousness. However, after some initial fashionable attention, we all just became confused and bored by it all. Then, once the economic crises of 2008 hit, all environmental commitments were pushed swiftly off the agenda, as if deemed null and void.

This deference to financial concerns is telling. This, if anything, reinforced my convictions; specifically, that the key to ensuring environmental responsibilities are met is to make it financially profitable. In 1987 the United Nation's Brundtland Commission defined sustainable development as: 'development that meets the needs of the present without compromising the ability of future generations to meet their own needs' (United Nations World Commission on Environment and Development, 1987). This concise statement says much about the self-determining aspirations of societies, as well as a long-term environmental responsibility.

It may seem obvious at this point to state that an energy solution this far-reaching must avoid environmental damage. But the projects with a proven commitment to environmental legacy are embarrassingly few in number; globally, we are causing more environmental damage now than we have ever done in the past. To continue in this vein risks denying the very future we seek for those we are trying to help.

Therefore, I will state it again here; what is more, I shall assert that *a product's environmental legacy contains commercial value in and of itself, and thereby a market opportunity*. Even when avoiding fossil fuels altogether, renewable technology solutions cause significant environmental impact from production, transportation and most critically, disposal. If PV lights containing electronic components and toxic batteries have an operational life of 2 years with no means of recycling (component separation or logistics), then there will be massive environmental damage at the scale of users we seek to support. More than 50 million PV lights have been sold over the last 10 years or so, and I know of no suppliers with adequate recycling schemes.

> A product's environmental legacy contains commercial
> value in and of itself, and thereby a market opportunity

The most realistic environmental policies set out to consider the materials used in the production of products. They then make those products last as long as reasonably possible to reduce the cycle of disposal, and include mechanisms for effective recycling products at the end of their lives. The next point is crucial. If the products are designed to be recycled – that is, if individual components and materials such as plastics and metals can be separated easily – then, at the scale we are considering, these collective items *have financial value*. We can use this as motivation for ensuring effective recycling of constituent parts for a high percentage of products in circulation.

Such considerations cost money. It will not surprise you that using recyclable materials is more expensive from a design and procurement perspective, as is designing products so that their components can be separated for recycling at the end of their life. Yet the markets we aim to serve are dominated by lowest cost purchasing, which is why hardly any products incorporate such features. Even utilising batteries that last 5 years instead of 2 (easily achieved in most PV applications) is not common for reasons of price alone.

Versatility and reliability

Long operational life is not particularly difficult to achieve in itself, and plenty of manufacturers are capable of achieving such longevity – if only the budget allowed for it. Similarly, the basic requirements of environmental resilience in product manufacture are a well understood discipline.

However, both long life and resilience become significant challenges when we consider the diversity of users and the environments in which they live. Such variance makes it particularly difficult to design and produce specific product solutions to meet all needs. The products and services we supply must be adaptable in physical scale and capacity; they must also be intuitively versatile in their operation to enable users to tailor functionality – value – to their own requirements. And all variants of our energy services must absolutely meet the challenge of reliability.

Achieving this is the key to making energy services attractive. The real and perceived value of consumer products lies in their reliability, compared with the available alternatives. As such it defines how much people will pay for them.

Take an obvious, large-scale example. Mains electricity systems in developing nations do not work effectively. There is every reason to believe that they will become less reliable in time, and little to suggest that they should improve. Providing reliable energy services, on the other hand, independent of utility electricity, is one of the core value propositions for our strategy. Against such a backdrop of deteriorating grid systems and fuel procurement, if we can offer high value to people who are unsatisfied with their grid or fuel-derived electricity, and make this reliability known to everyone else, then our customer base is only going to grow.

Versatility is a difficult thing to design. But *reliability* is more challenging still; it is closely dependent on the quality of manufacture, and, importantly, how the product is used. To rule out a common mistake, 'quality' does not equate to 'features' – loading products with functionality in order to meet a fantasy specification (or a focus group-friendly cost point, for that matter). Nor does it necessarily mean unrealistic build quality, as if trying to make products last forever. Prosaic as it might seem, 'quality' is about making a product adequately fulfil the task for which it is purchased, with resilience to the environment in which it will be used and a decent operational lifespan. This requires specialist skills. Energy-poor regions are the most demanding commercial markets of the world. The challenges – of environment, scale of geography and population, diversity of users, language and culture – all warrant an attention to design excellence, failsafe usability and dependable output.

Few organisations are capable of this scope and reliability. But they do exist – just not necessarily in the cheap mass-market consumer sector. If we remember, the true value of PV was initially demonstrated by industrial applications, by way of sturdy reliability and resilience in some of the harshest conditions on the planet. It is this sector which provides us with our benchmark.

The PV lighting sector demonstrates the limiting reality of the cost-competitive marketplace. Dozens of products are approved under the World Bank's Global Lighting initiative, or supplied under internationally funded programmes, and all fall short. Many have been developed by organisations who understand need – the environmental and user challenges to reliability and operational life. These organisations have their own staff based at manufacturing facilities (many of them in China) to oversee quality control; and they make significant efforts to educate users. But many of these products are not waterproof; nor are they rugged enough to be dropped on hard surfaces. Few are remote controlled, and hardly any have locking facilities or versatile means of mounting the lights. Plugs and sockets, push buttons, electrical connections and cable accessories – many requiring everyday use – regularly fail. If you incorporate valuable product features, you increase the challenges of quality control, while price pressures in the market make continuous product development (essential where rapidly advancing technologies are concerned) very difficult to support financially. Most credible products now offer a minimum of 2 years of operation, but few offer as many as 5 years. The result is a cycle of short-lived product dominance, of little more than 2 years, before being supplanted by a more modern version.

One supplier after another has tried to expand capacity, only to be thwarted by quality issues. The result is a product which provides value only in a limited set of territories, due to environmental challenges (such as a dry climate rather than a humid or wet one) or problems of localised training or support, where widely dispersed regions cannot fall back on word-of-mouth communication enjoyed in denser communities.

Engaging the technology giants of the world

Projects for the developing world require designers who, in addition to form, function and contemporary trends, understand the fundamental needs of people, and the environments in which those needs must be met. Converting these into real-life product specifications is a rare skill.

The same is true for manufacturers. Any given company could squeeze out cheap plastic gimmicky products in the tens of millions of units, but only a few dozen can produce high quality solutions that incorporate electronics, are sufficiently versatile and are rugged to ensure a good operational life. Fewer still can strike the magic balance and make these products appealing to consumers.

> Projects for the developing world require designers who
> understand the fundamental needs of people, and the
> environments in which those needs must be met

That said, the core technologies are widespread and the skills in using them well established – we do not necessarily need cutting-edge, ground breaking innovation here. The problem is that they are presently only used in 'controlled' conditions. For example, lithium batteries are employed in everything from smartphones to portable computers, but these devices are designed to be charged by stable mains electricity.

Stand-alone energy solutions have been in the most unforgiving environments of the world for over 30 years. These systems can either be a single electricity generation technology or a combination of several. And they have far higher statistical and practical reliability than utility grids. The requirement, then, is to adapt these technologies to our clearly documented and understood needs. What remains is the difficulty of justifying their initial cost against the value they will provide.

Let us continue to profile our target companies. The organisations proven to be capable of supplying tens of millions of products, with the appropriate versatility, reliability, ruggedness and design appeal, are high-volume manufacturers – multinational consumer product companies.

Now, I use the term 'multinational' in a generic sense. Smaller companies can design and specify products and visual materials, and have them manufactured by third parties. It does not need to be a single company with its own brand and supply infrastructure. (In fact, in the field of autonomous energy services, it is unlikely to be an established company. Supplying holistic energy services, as we have described them, is a disruptive strategy for the existing energy sector and, like many other innovative movements, it is most likely to come from outside of incumbent organisations.) When I talk of multinationals I am also referring to the ability to employ the best commercial resources in the world, such as product development, technology innovation, marketing, graphic design and brand development specialists. However, for the purposes of this text I shall refer to big multinationals because of the scale, competence and brand presence that they invoke in the mind. I find this way of thinking is useful to remind ourselves of the intimidating scale and diversity of the autonomous energy services sector.

So, far from the initial, daunting, demoralising brief which demands that we 'alleviate energy poverty and provide development opportunities' for which we must 'design and supply millions of products', we arrive at something which can be more easily framed in one's mind. In short, a manageable task is to attract high-volume commercial manufacturers. In particular, we are seeking these big players for their ability to fund product development, and for their proven aptitude at selling hundreds of millions of units.

The mother of invention

When I was younger the motto 'necessity is the mother of invention' was a popularly used mantra within engineering, indeed within society in general. The Second World War is often invoked as an example; its grave urgency gave birth to such technological achievements as radar (which enabled commercial flights), rockets (which were the basis for space exploration) and code breaking (the birth of the modern computer and the security used in online commerce). But can we claim that this phrase is still true, in light of the inequality we witness in the modern world? Or at best, have we redefined 'necessity'? For the last 40 years, the day-to-day suffering of over 2 billion people has been an acknowledged fact. Just think of the technologies we have invented, commercialised, refined and advanced in that time. So where are the major technological advancements to address the acute, urgent, glaring challenges of our species? What does 'necessity' mean in this context?

Here is one way to look at it. Where the fundamental 'necessities' for a basic quality of life (food, shelter, warmth) are viewed by civilisation as *commodities*, they are no longer real things; they have become abstractions – ideas of things – each with a commercial value or status. (Perhaps the people who need them also become 'unreal'; remember our endless sports stadiums, full of imaginary people we couldn't envisage?)

Where this happens, I believe that the instinct to amass such commodities – to withhold them, to profit from them and to wield them as agents of status or power – will prevail over any other. And I mean *every* other; be it kinship, loyalty, altruism, charitable works, even – and this one may be hard to swallow – the survival instinct. The act of providing these goods to another human being requires a strong motivation indeed. I argue, then, that 'necessity', the driver of progress, means brand leadership, commercial superiority and shareholder returns. Technological barriers are overcome if there's a sufficiently large financial reward at the end of it.

The instinct to amass such commodities – to withhold them, to profit from them and to wield them as agents of status or power – will prevail over any other

---

Such complex and expensive services as the internet, mobile communications and satellite TV have been developed in response to massive commercial potential. Personal computing has advanced to the point where our smartphones are technologically more capable than the wildest predictions of only 20 years ago. The relentless development of new features is driven, not by 'need', but by trends that the manufacturer has determined hold sufficient novelty value to sell the device to a minimum threshold of customers. This is why a new fingerprint recognition system or a speaking assistant are more likely to be developed than, say, an intuitive education platform with a battery life of 100 hours. All of these are possible: but some provide a quicker and more discernible route to a market and a profit.

The legal fights over patents between the likes of Apple, Samsung, Google and Microsoft attest to this. Sales, brand, innovation, diversified product range, intellectual property and stock price are all about making money. I am not bashing these big companies, or denying the good they and their technologies have done. But I am noting that they respond to a narrow sense of consumer demand which will foot the bill – in other words, the target demographic who will pay a premium for the latest, biggest, fastest, most feature-packed, show-off-to-the-world device.

This, ultimately, is what connects innovations such as driverless vehicles, virtual reality tech and the 'Internet of Things' with our own endeavours in the field of energy services. Technology aside, they all serve an emergent 'need' in the form of an emergent *market*; and in front of them stands an absolutely enormous commercial potential.

Now consider that the Apple iPhone 6 and 6 Plus sold over 10 million units within the first weekend of their launch; Apple has sold over 1 billion iPhones in total. In perspective, a PV-enabled light, capable of 6 hours of illumination per night, charging most USB devices and lasting 5 years with minimum maintenance, is a walk in the park to produce. We are literally comparing a state-of-the-art computer with a good torch.

The commercial challenge

These commercial organisations exist to make profits. In order to begin, they must have confidence in a long-term business opportunity. The money, as I've asserted before, is plentiful and available. Primarily it comes from the various financial mechanisms already operating in energy-poor nations. The challenge is to harness the funds already being spent on energy, fuel and, most importantly, those being spent on services that currently fail to deliver quality and value; then turning them to the benefit of local people. Meanwhile, the money that originates among the poor regions can be optimised, so that people are either spending less, or getting more for their money – more dependability, more quality and more value.

> Commercial organisations must have confidence in a
> long-term business opportunity in energy-poor regions

The low-cost myth

Despite the extravagance with which some wealthy people spend money, the biggest single issue contributing to lasting energy poverty stems from an obsession with spending as little of it as possible. The lowest-cost-wins attitude is all-pervasive, and it undermines efforts to motivate organisations and people. As far as our commercial strategy is concerned, it is our adversary.

We define much of the world around us in financial terms. This is the case personally and as collective purchasers and donors. Take note of the primary message of commercials on TV, printed advertising in billboards and media, and shops selling products of every kind. Above all else, you will see that the *price* takes precedence as the most promoted aspect of the product.

All of this seems obvious. The prevailing economic system is, it could be argued, the most successful in human history, with regards to wealth generation; and most of us would take for granted the idea of competition, with price as an arbiter of value. But this is true only where the wealth is manifest. As ideologies go, such a system is so powerful that to decry it would seem like heresy. But there is a downside. This tunnel-vision obsession with monetary cost, to the abandonment of quality, advantages, benefits or value, is a significant factor for the ecological mess and inequality in our world.

Quality and the low-cost mindset

How do I make this leap? To get an idea, remember what we said about quality and value. We defined quality as 'fitness for purpose'. Some products are more difficult to assess than others; but logically, most items that we purchase have a quality threshold below which the item does not adequately fulfil its function. Put simply, a market which competes on cost alone will at some stage exceed this threshold. Quality cannot exist below a certain cost. Cheap foods are a classic case, in which taste, nutritional content, shelf life and environmental impact are secondary to the purchase cost. The pay-off is obesity and a barrage of other global health issues.

Energy products and services, meanwhile, are difficult to assess without technical knowledge. This is trickier still amid the widespread misconceptions about the technology. In other words, they require effort – motivation – to understand the benefit. In the absence of knowledge, cost becomes the default method of assessment.

Competing on cost alone in the developing world can be prompted by the best of intentions. And it is not a measure of intelligence; hard-working consumers and illustrious academics alike share in reinforcing this way of thinking. But money defines so much of the development sector either directly or indirectly, that we do not even realise when our priorities have become messed up, our efforts counter productive.

This is a mindset problem. It is not found exclusively within the energy-poor markets themselves; rather, it is fuelled by those feeding into them. Despite the immense disposable spending resource of rich societies, it so often trumps the importance of actual need. For evidence, just take a look at the immense scale of low-cost warehouse stores around the world. Most are not known for their quality, and they show that the flood of cheap products is a feature of any modern economy. Ultimately it is people who determine what cost is most appropriate, without necessarily understanding why.

The mercy of others

By definition, the people you are trying to reach start off with very little money, or none at all. This is the essence of the 'poverty trap'. Commercial organisations, who have the power to address energy needs with any meaningful capacity, have at best a limited interest in 'low value' or 'high risk' markets. Anyone with limited spending ability, or unable to meet the capital cost of modern energy services at all – in other words, those who most need them – will be at the mercy of donors.

As I have outlined, energy services are complicated in the challenges they must address. For those who are motivated to try to help the poor there is a prevailing tendency to make products and solutions low cost (under the banner of 'more financially accessible') or to enter the fundraising arena – which is generally geared towards reaching the maximum number of people within a given budget.

What does this have to do with 'motivation'? Try looking at it this way. The myth of low cost is essentially a myth of consumer choice. After all, at the sharp end of need, why wouldn't anyone choose the cheapest option? The idea of consumerism as a liberating force is the subject of another discussion, and another book. What matters for now is that the myth of low cost is a thorn in the side of poor regions; and our strategy. It turns consumers into passive recipients of 'things' – products, gadgets and trinkets, which are, in the absence of user engagement, inappropriately low in value. Producers, meanwhile, naturally follow that market. I have revisited the notions of quality and value, at some length, in order to demonstrate a different commercial motivation, in which consumers become active participants in realising the benefits of the services they buy; and in which producers, realising the enormous potential of such a market, choose to innovate better solutions within it.

Alternatives to cost

But how else are we supposed to measure success in a market, if not by the cost of our products? Some non-profits use 'number of people reached' to promote their apparent success or effectiveness, and to obtain funding. But supplying a million people with products that break in the first few months is a shaky justification for receiving more support or kudos. Numbers reached; units sold; or expenditure in training; all create an often-false impression. The truth is that you can make numbers say anything you need them to say. It would seem that most existing suppliers of PV products do this.

There is no effective means of assessing performance of products in rural, developing world nations. Reliable, quick or comprehensive feedback or assessment mechanisms are simply not available. The landmasses are too great and the populations too diverse and disconnected.

Defining minimum quality

Generally speaking, non-profit organisations do not have the capacity to change the 'lowest-cost-wins' mindset of the market. What about the international sector? These influential, supposedly non-commercial organisations could have been in a position to establish minimum localised standards of quality and challenge the market prioritisation of cost. But they have not done so.

Many programmes are attempting to do this, with the World Bank's Lighting Global accreditation the most prominent. Without a doubt, it has helped the solar lighting sector, but it has not prevented the dominance of bad products or remedied the failure rate for approved products on the market. Elsewhere, organisations have not taken the initiative. But in practical terms, they represent the best chance of breaking the dependency on cost and raising the bars of quality and value. The difficulty arises when you try to arrive at a universal, international quality standard. Even were there a competent body committed to producing one, the diversity of requirements places serious obstacles. For example, the minimum environmental resilience requirements for coping with extreme dry heat are quite different from those of a cold and wet location. Placing an overarching requirement on product capabilities will raise costs unreasonably and needlessly for everyone.

Mindful of this diversity, a solution would be some sort of balance between universal quality and region-specific standards. As an analogy, the base specification of a car is universal, no matter where it is sold – minimum safety requirements, non-hazardous materials, emissions and so on. But more localised aspects, such as right-hand drive, the height of the lights from the ground and the position of the number plate, are tailored to the specific regional market.

Large commercial organisations, with a view to their brand value and continued growth, care about defining minimum functional standards of quality. And they have the means to take the lead in this regard. They are able to position products at market-appropriate cost, but with high enough quality; they can combine the power of the brand with innovative sales and marketing techniques, in order to reduce the focus on capital cost. We have already seen the example of Apple, whose products prove that quality can generate massive demand. I am convinced that the same can be done for autonomous energy service solutions. The key is to create demand on the back of proven quality, thereby providing high value to users, and a positive loop of motivation for producers and consumers alike.

How to sell quality autonomous energy services

The challenge of energy poverty, and of raising the base level of quality in the autonomous energy markets of the developing world can be stated in basic terms.

- Attract the most capable technology companies of the world by showing where financial profit can be made and sustained.
- Engage millions of local people to facilitate sales, support and commercial development – again, with financial incentive.
- Create demand for the high quality, versatile and therefore high value products described by this strategy.
- Propagate understanding of quality, value and potential to everyone. The higher the level of understanding in the general population, the easier these tasks will be.
- Use visual information and brand as the foundations for all the above.

I have outlined the need for commercial technology manufacturers to develop and supply high quality and high value autonomous energy service solutions for energy-poor regions of the world. I have argued the need for large numbers of local people to form the supply and support networks. I have also stated that financial profit is the way to motivate these groups to apply their capabilities and efforts. The way to generate the necessary financial incentive for both of these generic groups is to increase the scale of the customer base. Instead of assessing the price of competitor products or the spending capacity of existing customers, energy services must be defined and supplied against the value they represent to potential customers – and not just in the developing world.

These quality products will not be the cheapest, so innovative methods must be used to sell them.

There is presently limited commercial potential in energy-poor regions – largely because of low value products. I believe that by increasing the value of energy service solutions, we are able to grow the commercial markets, far larger and stronger by orders of magnitude than they are now. The remainder of this book is concerned with how that can be achieved.

PART FIVE

Implementing the Selling Daylight strategy

CHAPTER 18
The need for a long-term strategy

Abstract

This chapter describes how we can attract large-scale investment and commitment from multinational technology companies into the developing world energy sector.

We start by turning the discussion from one of risk to one of opportunity. If revenue potential is judged to be absolutely enormous in the medium-to-long term, then actions will be taken to establish a position in the fledgling market.

We are witnessing the quintessential 'game-changer' in the scale of billions of dollars – a gigantic market potential for any organisation positioned in the autonomous energy services sector; not merely an alternative, but preferential *to the utility grid.*

The real prize lies in being the first organisation to expand – not just within an application or a region, but also globally through the enormous range of energy services that are in demand.

To this end, we discuss why a long-term approach is needed and why diversification of products and commercial focus as early as possible is essential. We also look at why the capital cost of this strategy is justified.

Profit, pure and simple. This is the single undeniable incentive with which to attract the most capable companies in the world to the autonomous energy services sector. At present, energy-poor regions are considered 'high risk' by commercial organisations. Often, this is justified. Taking a business case to these organisations demands significantly more than it would otherwise for industrialised markets; more than merely describing a market for solar lights or phone chargers, for instance. Nevertheless, I wish to show over the next few chapters that commercial solutions are entirely realistic; and how to unlock this profit potential.

Making the commitment

To gain any sort of business response at all, let alone to provoke action, the commercial potential must be enormous in the first instance. It must be made manifestly clear and accessible; a business case which understands how to recognise the *actual* against the *perceived* pitfalls, along with the many opportunities that the market presents. It must also propose how to reduce risk and maximise opportunity.

http://dx.doi.org/10.3362/9781788530705.018

For this, we need to look further afield than the developing world. History holds numerous examples of technologies which demonstrate explosive growth at a particular period in time: electric lighting; refrigeration; air-conditioning; radio, telephones and television; personal computers and mobile phones. All of these items achieved a critical market breakthrough in the hundreds of millions of units, and became ubiquitous once they became convenient to use, affordable and available. Crucially, not all of these were new inventions, nor did they appear on the mass market overnight, out of a vacuum. Simply, a confluence of their maturity and the market meant that their time had come.

There is strong demand for these world-changing technologies in energy-poor regions; and enough money available to pay for them. It will be clear that many of these are the foundation to development; enabling light, communications and healthcare; essential to tackling basic poverty and promoting economic and social mobility. Now, with the world's resources online, and the internet providing complete educational, business and social services, we have never had a greater opportunity to reach so many with such efficiency.

An integrated PV system can achieve this using existing technology. It does not need ground-breaking, expensive research and development (such as a new microprocessor, a new type of PV cell or a new battery); indeed, it pays to utilise mature, proven technologies in order to minimise risk and enhance confidence with suppliers.

At present, energy-poor regions are considered 'high risk'
by commercial organisations. Often, this is justified

Business cases and risk

The challenge of attracting large multinationals to sell energy services in energy-poor regions is not just about highlighting existing opportunity. There has to be a long-term business case if the idea has a chance of taking flight at all. Organisations looking to sell products are well aware of contemporary demand, of revenue and profit potential, and of the existing size of the market. Their problem is *risk* – the uncertainty of commercial markets and economies which simply do not operate in the same way as in the industrialised world. The word 'risk' carries with it a negative connotation; after all it describes uncertainty in the face of possible obstacles – the 'unknown unknowns' of insurance premiums and defence secretaries. Yet the discipline of risk management is widespread precisely because we aim to incorporate it into our lives as a knowable quantity – as a tool, and as a legitimate factor in our decision making.

Now, creating risk assessment strategies for the developing world does not happen overnight – so unfamiliar, so vast and so diverse is the territory that such efforts are likely to take many years. However, better than trying to quantify or qualify risk, we have the practical and strategic means to respond to unforeseen problems. We do this by way of holistic design, versatile hardware, accessible visual media and indigenous support and supply networks. These also allow us to respond to sudden *opportunities*, and it is this very dynamism that we must aim to encourage in the commercial bodies we approach.

There are many ways to assess risk, either perceived or proven. Dedicated organisations publish risk tables and 'doing business' ratings for most countries of the world, in pursuit of some kind of objectivity; risks associated with logistics, the environment, disease, natural disasters, the people, governments, exchange rates, reputation and so on. Despite this, many large organisations poised to enter the energy services sector are not doing so because the focus on 'uncertainty' appeals to an innate cautiousness, while blinding them to the potential markets – a game of rock-paper-scissors, in which risk holds the advantage.

A more positive way of framing this is that there must be a sufficiently large *potential* market so that the projected gain outweighs the percieved risk. There must be a longer-term reason for the significant commitment required when entering a sector of this size and complexity. From a supplier's point of view, market potential is about more than unit sales. The commercial world knows that specific areas of demand exist (lighting and mobile phone charging being the most well understood).

But this is an incomplete picture. They need to understand that initially they have a cost-effective means to reach the 'low hanging fruit' potential sales, but can then generate more customers, achieve higher numbers of products sold per customer, sell greater value products per customer, and achieve greater profit per customer.

If we are to attract large-scale investment and commitment from multinational technology companies we will need, among other things:

- Strong revenue potential
- Confidence that it is possible, both technically and logistically
- An understanding of risk
- Market growth potential

Of these, that last point is the critical one. Even where the immediate profit potential appears underwhelming or the risk is considered high, if revenue potential is judged to be absolutely enormous in the medium-to-long term, then actions will be taken to establish a position in the fledgling market. Those actions will involve risk management planning.

In earlier chapters, we observed that energy-poor regions are characterised not only by the hundreds of millions lacking modern energy services, but by an equally large number who depend on unreliable or deteriorating utility grids. Consider this in context with the increasing importance of electrical energy services in our world. Even if we set aside (for a moment) the opportunities surrounding those with reliable networked grids, we are witnessing the quintessential 'game-changer' in the scale of billions of dollars – a gigantic market potential for any organisation positioned in the autonomous energy services sector; not merely an alternative, but *preferential* to the utility grid.

Feasibility and confidence

It is entirely possible that a large brand name will emerge as a dominant supplier of autonomous energy services over the next decade. In fact, it is perfectly probable. How shall we approach this opportunity?

I believe that the strongest driver for change in energy-poor regions, and the correct basis from which to address all subsequent issues of development, is *the potential for making money*. I will not claim that it is easy to make giant profits, nor that it will happen overnight; instead, as described above, I shall invoke the long-term strategy, approaching the opportunity in order to maximise business potential, because this encompasses the sources of revenue, motivation to deal with risk (instead of allowing it to prevent action), market growth potential and logistics. Let us revisit the principles of this plan as we have established it so far, revisiting the technical basis and feasibility for autonomous energy.

- Sunlight is everywhere and PV captures and converts it to electricity. The stand-alone PV format allows that electrical energy to be stored and utilised as required. These stand-alone PV energy systems can be applied to all manner of applications around the globe, across the entire potential energy services market. PV is the only technology solution capable of this.
- The 'holistic' technical approach of this strategy ensures high quality, high value and highly versatile products.
- A Visual Instruction System (VIS) allows us to reach vast numbers of customers who, until now, have remained at the mercy of local sellers in a lowest-cost-wins market. We can engage with users and enable them to gain more from our products.
- Critical from a commercial perspective, we enable the foundation of a unified brand identity in markets where it is notoriously difficult to become established.
- Developing confidence in our brand (against a backdrop of huge desire for electrical services) will create demand for our solutions. Once there is demand, market forces work in favour of the supplier, not against it.

So far, so good. But nothing is that simple. Remember that PV, with all its versatility and adaptability, carries baggage – namely the poor reputation surrounding its generic stand-alone technology. Any strategy involving PV will have to overcome this, which means differentiating products and solutions from anything tainted with the present-day perception of 'solar'. (This is one of the reasons I refer to 'energy services' rather than using 'solar' or 'PV'.)

We are witnessing the quintessential 'game-changer' in the scale of billions of dollars – a gigantic market potential for any organisation positioned in the autonomous energy services sector, preferential to the utility grid

Quality solutions and their cost

All of this costs money. It calls for greater capital expenditure than alternative approaches. What justifies this upfront cost is the value it holds for those with greater, longer-term business aspirations. This strategy – 'Personalised Energy', a visual instruction system, holistic technical solutions, knowledge sharing and branding – provides enormous long-term commercial opportunity in a vast market which is only going to grow.

The success of such a strategy depends on products and systems of high quality and high value. Designing and developing even the simplest end product, such as a solar light, can run into the hundreds of thousands of dollars. Of course, this could be achieved for less, but always to the detriment of the product's capabilities, and what it enables its user to achieve.

There may well be situations where we achieve a cost advantage through energy efficiency and integrated design, by way of our 'holistic' approach. But this is a temporary win at best. Familiarity with the market tells us that we will not be the cheapest forever. Copy products will inevitably come along to undercut us, deliberately designed to look similar and targeted directly at the same distributors and third parties. Suppliers will make competitive performance claims, regardless of their products' ability to live up to them.

But there is a window of opportunity. Our holistic solutions capitalise on the fact that lower energy requirements mean lower cost. They can therefore offer higher value and still compete commercially with present alternatives. I shall discuss the importance of this initial cost advantage later. But keep in mind that any cost advantage should only be a supporting factor in our strategy, not central to it.

Taking the long view

When any large company invests into a sector of this scale, there is more at stake than shipping a product and making a quick return. It is a long-term business commitment with significant capital expenditure. I am not suggesting spending more for its own sake; rather I am stating that that spending should be based on the long-term principles outlined here, ensuring reliability, longevity and stability, rather than throwing money at quick-return, short-term marketing strategies. Indeed, success in this arena depends on avoiding the short-term view. PV lighting product suppliers all know, to their cost, how to fall victim to one's own initial limited success. Certainly, in the first instance, quick sales targets will propagate the brand. But the more success one demonstrates in a restricted market, the greater the desire of others to compete for a share of it. Suddenly, the single market sector, or the limited application, or operating in a limited geographical region, are not enough. There is a need to expand – not just within an application or a region, but also globally through the enormous range of energy services that are in demand. The real prize lies in being the first organisation to do this, as it affords the ability to define the industry standards to which all others must compete. From a supplier perspective, there are smart ways to go about this.

This strategy provides enormous long-term commercial
opportunity in a vast market which is only going to grow

One product, one application

A single product, with a particular application in a specific market (and drawing on one type of budget) can be developed into multiple products, with value in a range of applications and markets, financed from a range of budgets. This is the classic expansion model, and it is the business case large multinationals need – from selling solar lighting products and mobile phone chargers to powering commercial infrastructure projects such as banking, phone networks and public services for entire regions. Once a brand is recognised as a service supplier of merit, the national programmes for energy – the ones presently supplied through massively wasteful procurement processes – can be attracted and fulfilled by more cost-effective, better quality service solutions.

For a moment, consider a simplified objective. Let us say that we wish governments to spend $100 million on autonomous energy services (instead of, say, a coal-fuelled power station which feeds utility spurs of limited geographical reach). The confidence required to award this scale of contract to a distributed energy provider will take some time to develop. The initial cost advantages for high value products from this strategy will help, as long as there is an understanding of the undeniable commercial value of the approach I am describing.

This strategy, then, calls for more than accessing the funding sources and budgets. It is also about being smart when meeting these costs over the elongated timescales necessary for success.

Investors waiting for certainty that will not come

The autonomous energy sector needs substantial commercial commitment, and a lack of investment proves a fundamental barrier. At the outset of this book, I identified four principal types of people with respect to their varying access to electricity – from reliable, mainly urban industrialised utility grids, to rural poor with no access at all, and a vast spectrum of unreliability in-between. The autonomous energy markets have now reached critical mass in terms of commercial demand, because what unites these customer groups is their need for a dependable service.

So the problem is not one of demand; rather, it lies with the self-appointed sector suppliers, such as the PV industry; it concerns people, and what they consider autonomous energy can do; but most of all, it lies in knowing an opportunity when we see one. There are no exact numbers to express the potential customers in our sights, their spending power or the overall financial opportunity for what we are describing. The sheer scale and diversity of the developing world preclude this.

The financial sector is keenly observant of the developing world (more specifically what it calls 'emerging markets'). The trouble is that they are waiting for the hard numbers: profit potential, timescales, risk factors, market size, SWOT analysis and the rest. You cannot poll all of the people with money to spend in energy-poor regions. Many have no formal addresses; some don't have formal names. You can't analyse their present expenditure as you would in the industrialised world. People are not part of a commercial online market. Hundreds of millions of people buy energy every day; these people do not have a bank account. And despite enormous user growth rates over recent years, not everyone owns a mobile phone.

Where data does exist, it can mislead; or in the case of figures put forward by international agencies, it can simply be out of date. A 2012 report on earnings per day (for example) may draw from data gathered before the global economic crash. Transactions are largely based on cash or goods. The spending is definitely there, happening in the tens of billions of dollars per year. But quantifying it accurately is, in practical terms, impossible.

So for our analysts, the data – the certainty – is not forthcoming. Those hard numbers won't appear before the opportunity is addressed; they will come afterwards. And in the event, those same hard numbers will be reporting the success of the company that went in first and did it right.

Estimates of market value

For the time being, then, some estimates. I take the same approach with the value of the autonomous energy market as I do with the scale of the need: look at the magnitude of the numbers instead of thinking of the specifics. First, take a few sources of market potential (all in US dollars): as early as 2007, the World Resources Institute predicted a US$433 billion market for energy in its paper 'The Next 4 Billion' (Hammond et al. 2007); the UN Environment Programme estimated global savings for off-grid lighting, where solar-powered LED lanterns displace kerosene and candles, totals somewhere between $25 billion and $33 billion. Bloomberg N. E. Finance (Bloomberg, 2012) estimated a $660 billion market in diesel replacement by PV (if off-grid PV was $1/Wp. It is now below that cost.)[1] From GSMA figures (GSMA Intelligence, 2015) it is estimated that $40 billion is spent just charging phones – for which PV is the most versatile and economic solution. Elsewhere there are various estimates of $30 billion+ spent on disposable batteries; even higher estimates regarding spending on secondary access to mains electricity, for charging devices or internet access.

You have to assume a great deal of variation in these numbers, because the data is hard to gather and so many assumptions need to be made in presenting them. Some of the estimates could be as much as 50% out either way in terms of market potential. But while the financial institutions keep their analysts sat staring at the numbers, they are overlooking the sheer diversity of spending that can be accountable to modern energy provision. The only numbers likely to catch their attention in their pursuit of certainty will be the annual report of a trailblazing commercial organisation, showing details of how many millions profit they made the previous year. The next sound you hear will be the bandwagon creaking as everyone jumps aboard.

The opportunity for the profitable sale of autonomous energy services, supplied in the hundreds of millions of units, has already arisen. There is a critical mass and a critical movement of demand.

The need to diversify

One particular obstacle for large organisations entering the energy services market is an over-reliance on the standard business ways of thinking. A common approach is to focus on specifics: find one opportunity and solve it better than anyone else. But this does not work in the developing world energy sector. I do not believe it is possible to grow, less still sustain, a business in just a single sector. A supplier of energy service solutions needs to diversify – and do it as quickly as possible.

PV lighting product companies continually demonstrate this. The company that sells a portable PV light to an individual should never have to refer that hard-earned customer to a competitor when they are ready to purchase lighting for their home. When patrons of a local bar develop a preference for cold beers and sodas (as opposed to the usual warm ones) courtesy of a PV-enabled refrigerator, they should be able to source a fridge solution for their home, office or business from the same supplier to the bar. The same goes for the people who visit PV-enabled internet cafes, mobile phone charging facilities, air-conditioned rooms and so on.

This informal expansion of business opportunities is essential. While you would not base the whole future of the business on this type of reputational sales, it still needs to be a part of the growth strategy, and incorporated into risk planning. This is nothing new. It is the same reason we see multinationals diversifying into related products for their affluent market. Think of the congruent product ranges from organisations such as Sony, Panasonic, Sharp, Sanyo, Philips, Bosch, Samsung, Hitachi and others. However, traditional brand propagation isn't an option for us. Remember for a moment the daunting scale of the territory, and the diversity of the people we want to reach. The majority do not have television, so brand development and sales through TV advertisement is not a strong enough promotional tool. Radio is everywhere but it is language specific, making brand recognition (in terms of logo, colour, features and reputation) all the more difficult to convey. Billboards and fixed advertising are widespread, but they are inert and inflexible. Internet advertising will not yet have sufficient reach. As if this were not enough, even your current customers remain prey to the lowest cost option of competitors.

The focus and reach of established product and brand promotion in the industrialised world cannot be replicated in the developing world. Nor, for that matter, can the convenience of product availability. Nonetheless we need to create demand, and in the absence of omnipresent PR and marketing tools, that takes time. A single market sector is not sufficient to sustain commercial growth – especially with the continuous development cycles essential to maintain a strong product.

The need for a diversified business approach is another reason why this strategy is so capital-intensive in the early years compared with more traditional approaches. It needs to be this way. Traditional business approaches are not satisfying demand. Considering that this is the market most supported by international agencies and the media, the most successful PV lighting supplier in terms of units sold has supplied, at a generous estimate, 10 million lights over a period in excess of 8 years.

The opportunity for the profitable sale of autonomous energy services, supplied in the hundreds of millions of units, has already arisen. There is a critical mass and a critical movement of demand

---

That is far less than 1% of demand. In fact, total PV light sales over the last 5 years, from all companies, barely meet 2% of the 1.3 billion people still using kerosene, candles or disposable batteries for lighting. PV lighting has international backing and hundreds of suppliers; its value propositions are very strong, as are the health and safety benefits. Yet 2% of demand is the result, to date. (It could be argued that many more millions of PV lights have been supplied but not counted within official figures. But a good 50% of the lights supplied will operate for less than 3 years – many for less than 1 year. So in terms of serving people in need the 2% figure is representative of those that actually have a functioning PV light which negates the need for fuel-based spending.)

I have approached many interested parties with a view to implementing this strategy. Most have wanted to start by concentrating on a few strong areas of demand. They normally do a bit of research then come back wanting to make a solar light, a phone charger, a hybrid of both or, more recently, a solar home system. None of these people is involved at first hand with the realities of the developing world, where a one-track approach puts them at an immediate disadvantage.

A good company with a good product struggles to maintain business, let alone grow, in the developing world. Why does this happen? I have had numerous discussions around this subject with people of all disciplines; trade books and commentators likewise throw some light on the issue (excuse the pun). Some of these reasons will be familiar. The pervasive focus on lowest cost plays a part, for sure. Money is hard earned in these regions and the concept of disposable income is alien to many. Unfortunately, so are the concepts of quality. Our old favourite – the troubled reputation of generic PV – is another definite factor, though quantifying its impact is impossible. Next are those factors arising from individual attitudes, regional preferences and cultural short-sightedness; which, while entirely counter productive, still persist. Meanwhile, environmental concerns are largely non-existent, as is customer loyalty.

A trader can buy from the same supplier for years, enjoying favourable payment terms, a good working relationship and an understanding of their business needs such as volumes, packaging sizes or delivery schedules. Then a big order comes along and the trader will go to a competitor for a 5% purchase saving. The new supplier will mess them around: supply faulty goods, demand payment upfront, define supply terms in their own favour and generally be unwilling to meet any of the trader's requirements. The deal will cost the trader more overall, but the faulty logic is all concentrated on the singular purchase cost. The original relationship is broken and a new one does not form. The damage is done.

This is an endless cycle commonly referenced in many sectors and at many scales of commerce; and governments and aid agencies do the same thing. A local specialist off-grid PV company will be approached for free advice on implementing energy services in remote areas. Following a conscientious recommendation, sometimes a small pilot order and unbudgeted time, things go quiet as an overseas competitor wins the contract. When the systems fail, the local company is contacted, for advice. We have already seen the formal procurement challenges in an earlier chapter. But it is the dysfunctional mechanisms surrounding them which prolong such a frustrating cycle of decreasing value. Corruption, of course, exists; staff are not qualified for their positions; and bureaucracy at every stage threatens to derail the best of efforts.

What is my answer to all of this? For a start, deal with people who have already worked out how to navigate their small corner of the commercial landscape. Incentivise them financially (and legitimately) and have them, and millions of others, work *for* the supplier rather than against them. More on this shortly. But the overall message is: don't put all your eggs in one basket. If all the commercial efforts of an organisation are focused on a single product, or a few products that collectively are only valuable for limited applications or regional markets, then profits will be difficult to obtain, and medium to long-term growth seemingly always out of reach. The answer is for the organisation and its products to be versatile, adaptable and resilient within the global energy services sector.

Endnote

1. Watt-peak, or the maximum power output of a 1 square metre solar panel at 25 degrees centigrade.

CHAPTER 19

Why and how to be a diverse company

Abstract

Here we discuss how to make the hardware of our stand-alone PV solutions as versatile as possible.

Versatility in function and form factor can be utilised to provide significant commercial value. This can encompass profitable end-of-life recycling, addressing multiple sectors from a single core product, being able to bridge global markets and responding to unforeseen opportunities.

Using the example of autonomous energy services for tents, we demonstrate a single product concept that can offer multiple applications in multiple markets, drawing on multiple budgets and reaching multiple customer sectors.

Yes, such versatility can be costly, time consuming and difficult to develop, but the end value of such an approach is truly immense. Financially, this effort is justified against confidence in high-volume sales.

Once we are able to look at the whole autonomous energy sector from a position of high quality and high versatility (while still being cost competitive), we can develop confidence in the truly enormous scale of the commercial possibility.

Stand-alone PV can be described generically as 'autonomous battery chargers that operate at zero cost'. The sun is the most ubiquitous energy source; PV is the most effective electrical solar technology; and stand-alone PV is the most versatile energy services format. These systems benefit from inherent reliability and widely available core components. They are environmentally clean, silent in operation and offer proven long life. The value of 'Personalised Energy', as a strategy, is rooted in long-term commercial potential. Part of that value lies in its versatile, holistic technical approach, in its Visual Instruction System for sharing knowledge and in the brand.

PV as a consumer product format

But maybe the most attractive feature of stand-alone PV systems is their ability to be supplied and purchased in the form of a consumer product; which basically means that the whole solution, including instructions for assembly, installation and care, can be supplied in a box, with no need for additional resource. This format can meet the needs of the highest number of people. It allows for widespread distribution utilising existing supply and support

http://dx.doi.org/10.3362/9781788530705.019

networks – international, national, urban and rural. It is also the only mechanism which is proven to be effective at the massive scale needed.

The problem is that stand-alone PV is generally not designed or supplied in a suitable consumer product format at the moment. Manufacturers face a formidable requirement – to adapt the formal design principles of stand-alone PV to the well-understood methods of consumer product manufacture.

Maybe the most attractive feature of stand-alone PV systems is their ability to be supplied and purchased in the form of a consumer product

---

Maximising potential

Launching a diverse range of energy service solutions requires us to maximise market potential, if it is to be commercially profitable. Critical to this is optimising both the response time and the cost of meeting the opportunities as they arise, all the while ensuring and safeguarding quality.

Our Visual Instruction System allows value propositions to be conveyed across all regions of the world, to all potential customers. Once produced, it is ready for deployment at a moment's notice; no translations or regional adjustments are required (although materials can be tailored for a specific opportunity if the opportunity justifies it). Brand recognition will also help grow confidence in our ability to supply solutions for the next application or the next customer. But there must be an ability to supply products that meet demand quickly and cost-effectively; and for this we need versatile hardware.

Common core technologies

Electrical products, particularly those considered 'smart' technologies, can be thought of as having several distinct sections. This distinction goes further than mere 'software' and 'hardware'; at a deeper, conceptual level of design, we can identify central, discrete, core elements – modules if you like – which control a range of their principal features and functions.

What does this mean, and what serves as an example? In the case of stand-alone PV systems, the charge controller is a distinct unit, whose primary function is to maximise battery performance – by controlling the input and output energy of the battery. It measures the input and output voltages and currents, as well as temperature, humidity and other properties; it then performs calculations to determine the most appropriate mode of control. But the actual control itself – the switching or manipulation of the charge energy and the electrical output from the battery – is performed by discrete electrical components. The core algorithms of the controller can be performed by a microprocessor no bigger than a fingernail (for the cost of a dollar or so) yet the controlled energy levels can be anything from microwatts (as in a calculator) to kilowatts (a telecoms repeater station). The brains in control simply manipulate the devices appropriate to their power needs. The same core control system can be used in hundreds of outwardly different applications and applied to any scale of solution.

This principle of a single core system controlling a range of optional peripherals is widely practised across industry – and not just for controllers or their subordinate components. Entire technical solutions can simply be employed within differing housings for branding or other purposes. This is true for air-conditioners, microwave ovens, kettles, TVs, computers, mobile phones, washing machines, in fact pretty much any appliance you care to name.

Another standard approach is to employ the same core components to perform the essential functions, and include or omit certain user-facing facilities according to price, regional need, customer demographic and so on. This is why the control surfaces of many products have blanked out spaces where you would otherwise expect to see additional switches or indicators in the top-of-the-range model. Examples would be an optical input in a top-of-the-line stereo; an internet input connector on a TV; a switch for the heated seats in the luxury variant car – all occupying the blank insert displayed on the budget model.

Designing such feature-rich core controls of course proves more expensive than a device with the minimum of essential features. But once the development is complete, the final core control system becomes incredibly versatile – and therefore valuable in a whole range of applications and market sectors.

An example of core versatility: tent applications

The commercial application of tents provides a good example of the value of core versatility. A basic energy service product for a tent comprises a flexible PV panel, a battery pack and a control unit, and is packaged with the necessary cables and fixings. The control unit, in addition to its high-level functions, includes two USB sockets. This most basic product is a mobile phone or other USB device charger (which doesn't have to be sold just for tent applications). It includes the standard visual literature for correct usage and care: how to mount the PV on a tent; how to optimise charging with respect to the sun; keeping the PV clean; handling guidance; and keeping the battery pack cool.

This basic product, intended for personal or domestic use in energy-poor regions, also has value for industrialised world applications – where a user may require remote charging capability while at the beach (PV mounted on a sunshade), hiking (PV on a backpack or wind shelter) or camping at music festivals. Next, add a small LED light to the package and its value increases. The product now has appeal for recreational camping, but also for emergency shelters and refugee camps, where millions of tents are deployed each year.

Autonomous energy services for tents demonstrate a single product concept that can offer multiple applications in multiple markets

Adding more PV capability (as an optional extra) enables functionality in less sunny locations or poorer weather; additional batteries allow more devices to be charged and brighter or longer lasting lighting. This give us the next product in the range. In good sunlight conditions this has sufficient power for a movement sensor and alarm so that a security function is enabled for the tent. For the music festival market, a remote control can make the security-enabled tent flash its lights – obvious value to anyone who has returned, somewhat worse for wear, to a field of 300 tents at 3:00 in the morning.

Using the same core controller, we can make the whole system larger still to power a fridge. Again, this has a range of applications – a cold drink, baby products, perishable foods, medical vaccines, chocolate, wine, water and so on. Each application has commercial value for a range of markets: field vaccination tents, disaster and emergency response, recreational camping, professional outdoor pursuits, military, sporting events, medical tents or just beach parties. Larger systems still (again, based around the same core controller but with higher capacity peripherals) enable further lighting, fans, TVs, projectors or a whole range of functions for marquees, hospitality tents, garden parties, wedding parties and sponsored areas at sporting events. The controller includes data collection as standard and is compatible with a range of user interfaces. Ultimately, the core technology at the heart of our humble tent-based charger can be turned to specialist applications: such as inflatable tents that are deployed by disaster relief agencies, particularly in flooded areas; or alternatively in mobile cinemas, for education and entertainment in unelectrified regions.

Autonomous energy services for tents demonstrate a single product concept that can offer multiple applications in multiple markets, drawing on multiple budgets and reaching multiple customer sectors. Although the development of the core technology is more expensive than just developing a 'PV light that charges a phone', the resultant versatility returns huge value. If you think the tent example above exhausts the list of possibilities, consider further applications which fit this model:

- Refrigeration: for vaccines; household food and drink; bars and restaurants; distribution centres; and cold storage chains from portable fridges to cold storage warehouses.
- Communications: for personal phones; households, small businesses, private networks (banks, retail stores, government offices); and national infrastructure projects.
- Entertainment: for smartphones, TV, computers, projectors and cinemas.

Remember, stand-alone PV systems are silent and environmentally friendly; they require no fuel and have no moving parts; they can be assembled by the owner; and, if the visual information is followed correctly, they are highly reliable, or at least predictable and controllable.

The value of microprocessors

Microprocessors are truly incredible things. They add functionality to thousands of items in our lives, often without us even realising. For our purposes, they can perform all sorts of tasks such as monitoring, control, logical decision making, comparisons and status reporting. As the tent example demonstrates, microprocessors provide more than mere features. They unlock clear commercial value in a range of sectors – and therefore enable sales from a range of funding sources. This is critical for commercial risk mitigation.

We were talking about maximising potential. How do we realise the commercial value of a PV system, and what does this mean at the level of our core components? At the outset, we should expect a core microprocessor unit to boast the following features.

- Options for extending battery life for applications which have to last many years. (This is at the cost of reducing the daily available capacity; and is the same principle of performance-versus-battery-life employed on laptop computers.)
- The facility to optimise different battery technologies, which exhibit unique responses to different electrical and environmental conditions.
- Control options which are selectable (or automatic) to cope with particularly high or low ambient temperatures.
- Variable user interface options.
- Different sensor options, such as movement sensors which efficiently control lighting; or an audible alarm that sounds if a fridge door is left open (detectable by the controller in the form of excessive energy drain).
- PIN entry for security, payment option and rental of units.
- Magnetic sensors for simple testing.
- Diagnostic readouts for more complicated fault or performance analysis.

Physical versatility

Versatile design is about more than functionality, or extensible features. It must also take into account the *physical* aspects of a product or solution, including the environments in which it is expected to be used.

This is an expensive business, even before you reach the point where you can begin manufacturing products. Plastic or metal product 'bodies', for example, require expensive tooling (tens of thousands of dollars for even a basic plastic moulding or metal extrusion), so it is important to get the design right before committing to a single, adaptable base component type. External colour and component shape options are valuable; but so too is the fundamental ruggedness of the housing, the water resilience, and accommodating local versatility for items such as switches and connectors.

To add a further layer of challenge in the interests of versatility, all of these options should be developed for products of many appearances. The internal structure of the product may be universal (for supporting the core controller for example), but the option to employ a range of components or body parts should be incorporated. An example is car production, where a range of cars from different manufacturers can actually share the same running gear – from the wheels to the engine and gearbox. A different 'shell' (the body, dashboard, seats and other parts) is selected depending on the commercial specification.

Next we consider a product's environmental credibility. Having a common core with a range of optional discrete components naturally lends itself to end-of-life recycling, because all of the major components, by definition, can easily be disassociated from one another. Sadly, environmental considerations hold little value, not only for customers but also for many manufacturers (including many 'sustainable' energy solutions). But the nice thing about designing versatility into products for profit-making reasons is that it naturally enables ethical disposal. The core systems are separate from the plastic housings, the battery, metal mounting brackets or other discrete items. A modicum of consideration at the DFMA stage (design for manufacture and assembly) will make it easier for these items to be separated from the product at the end of its life, and appropriately disposed of or recycled. Simple techniques include allowing separation of value items such as batteries into the preferred form of battery recycling companies; and indenting the specific plastic type into the actual moulding process – for permanent identification and categorisation for recycling.

The value of versatile development

Products for the developing world do not see anything like the capital commitment to development that they merit. A pressing concern is the level of sales confidence in a market sector that is known for its unpredictability. Another, of course, is price sensitivity. The result is a compromise in product form factor, and in the capability of the core product to meet new requirements and sales opportunities.

Yes, such versatility is costly, time consuming and difficult to develop; and the cost of mistakes is high. (3D printing and computer modelling are going some way to reduce these costs.) However, the end value of such an approach is truly immense. Financially, this effort is justified against confidence in high-volume sales. The costs will be amortised against the first, say, 500,000 units. Thereafter, every unit sold brings in a higher margin of profit.

Responding to unforeseen opportunity

Core microprocessor circuits, and their associated software, should be designed to perform a comprehensive range of functions. In many products, these will not be utilised; but once the functional capabilities are there, it is simple to include them if a new commercial opportunity arises – instead of taking months or even years to develop a specific new product with a new set of capabilities.

Versatile core technology gives you the ability to adapt to such opportunities, and to meet the requirements of regional populations. It is not uncommon for these to surprise even well-integrated regional suppliers. You cannot hope to get a straightforward list of requirements from over a billion people, after all. Having developed charge controllers for stand-alone PV systems, I appreciate the time, cost and resource required. But the experience has shown me just how valuable versatile control systems really are. Commercial opportunity can arise seemingly out of nowhere; it leaves suppliers scrabbling to put together rapid compromise solutions, but ultimately missing the opportunity. Be prepared.

Versatility against unforeseen challenges

Throughout this chapter I have stated that upfront costs are higher for this strategy than they would be for selling a product in specific markets and 'planning' – hoping – for 'organic' or controlled growth.

Another strong reason for going the long way around is that you cannot analyse the market with any kind of certainty. Even if you know the scale of the PV lighting market, for example, you simply cannot predict the problems you are going to encounter – particularly the ones that you should not have to face.

One such problem, and a serious cost to many suppliers, is battery degradation. Products can be held or delayed by customs for months on end for various reasons. All the while the batteries are draining in the cheaper products. Ironically this degradation is exacerbated by the very sunlight that would otherwise be charging them; instead, the intense rays of energy are hitting the metal shipping containers holding the products in an already hot environment. This long-term storage problem could be avoided if quality batteries are used and the core controller features a sleep mode to minimise battery drainage.

Elsewhere, we see a multitude of commercial and political challenges which prevent businesses growing in any controlled manner. Different barriers arise across a single country, let alone from one country to another. This should come as no surprise: as explained earlier, it is the diversity of potential customers that creates the problems, not the numbers. This diversity encompasses business dealings, dependability of individuals, issues of trust and professionalism, and, not least, competency.

The nature of the developing world means that it resists conventional corporate planning. There is no ideal pre-defined order in which one chosen application is followed by a second, planned by people in suits sat around a polished table. Nor will one country necessarily take precedence over another in a planning sense. We know that tens of billions of dollars are being spent, and that this spending is incredibly difficult to define. Rather than trying to analyse the specifics of where to position a product or offering (this will only be needed when deciding where to start), the trick is to conceive a flexible business strategy, whose versatile products capture opportunity across the whole market.

The most effective way to minimise risk and maximise commercial potential is to accept the unpredictable nature of the regions in which we do business, and to approach the overall autonomous energy services market as a whole. It is to ensure flexibility in the business and versatility in the products to adapt to opportunity as it arises. Because this is precisely what happens: very large opportunities will arise without much notice, and you don't have the luxury of a year to develop a quality product to meet it.

A large national project may suddenly be announced with little warning (or preparation), sometimes calling for you to deliver within six months or less. The reasons can vary: a budget may have to be spent within a financial year; a crop may have failed; or a political issue has been raised on the international stage and there is sudden impetus to be seen to act. Typically, these tend not to be awarded for the most suitable product, because often none of the suppliers capable of the volumes and the timescales *has* a suitable product. Contracts go to the organisation in the right place at the right time, or with the right political connections.

Responses to natural disasters are a bleak testament to this process. The world reacts with great sympathy for a few days, donating large sums of money before retreating to its familiar lifestyle distractions. The money needs spending quickly, and in the absence of appropriate energy solutions, the selection exercise becomes a scrum for the attention of the response organisations. Back at the sharp end of things, outside of acute emergency aid such as food and medicine drops (and I am careful to make the distinction), those suffering in the aftermath of the disaster face being the early recipients of cheap 'solar lights' or phone chargers which may only last a few weeks or months. They themselves will be categorised as having been 'supplied'; this box ticked, they will have to wait many years before being seen as eligible for 'further' or lasting assistance.

What are the options? You don't have the time to develop a product from scratch. Then again, if, as is the case for most suppliers, you rush to adapt an existing product to the needs of a given project, you risk undermining its quality and, by extension, its suitability or value. Ultimately, as is typical of these reactionary, panic ventures, your profits will suffer when product failures start to occur.

The market suffers accordingly. When inadequate products with the apparent potential for giant volumes (in the hundreds of millions) experience only modest sales, the poor showing in low volumes and low margins does not go unobserved; and any investment in development becomes restricted. In a familiar chicken-and-egg situation, the market is blamed as insufficient to justify the development costs, rather than the lack of development being recognised as the cause of the small commercial market size. Put bluntly: the stand-alone energy services sector is being stifled by existing products and suppliers.

But there is a third way. If you have versatility in your core product from the outset, by virtue of its central microprocessor functionality, and if you assemble it wisely as discrete peripheral items and with, say, generic but adaptable plastic housings, then you have the means to submit a quality option.

It is the smaller products which benefit the most from this functional and physical versatility; the personal chargers, lights or basic solar home systems – as they tend to lack the budgets for incorporating extended features otherwise enjoyed by the larger energy service solutions. And with a shared core commonality with their larger cousins, we get high functionality and broad appeal for small, relatively low-cost energy services. This is critical to market development – and for combatting energy poverty. The two are not mutually exclusive, as we shall see.

Put bluntly: the stand-alone energy services sector is
being stifled by existing products and suppliers

I maintain, then, that the market reputation hinges crucially on *improving the value of small products*; the ones with which suppliers feel most confident to try to sell. And it is the small product markets, such as lighting and mobile phone charging, which define opinions on the overall potential of the technology, and of the industry.

Bridging global markets

In our tent example from earlier, the core components generated revenue, not just from sectors within the same regional populations, but also from different demographics. Wealthy beachgoers purchase mobile phone chargers with disposable income, as do festivalgoers. In dozens of applications, the common functionality can be packaged to industrialised markets as well as developing ones.

Bridging these markets is critical. Versatility is plainly about more than squeezing as many applications as possible from a core technology. It asks us to understand the technical and market value for energy services in *all* societies, not just those with energy challenges. In basic terms: modest volumes in the industrialised world yield strong margins; they support the relatively low margins but high volumes encountered in the developing world. Manufacturing becomes cheaper per unit with volume. But it is the development stage which benefits most, thanks to a boost in confidence for investing the capital, time and resource. Industrialised markets are simply better understood and perceived as less risky. (Although, to my mind, fickle buyer trends are just as unpredictable as developing world markets.)

The benefits of being able to sell what would otherwise be 'a developing world product' into industrialised world markets are strong indeed. How tangible is this, in reality? It's debatable. In my work with autonomous PV solutions, it has always helped to be able to sell into markets which boardroom members can relate to first hand; even if, realistically, the commercial profit potential is tiny against that of the developing world.

As I made clear earlier, industrialised regions have mains electricity everywhere, with various options for portability when not physically connected to the wall. Stand-alone energy services have only limited value, and then for a very small percentage of society. But in energy-poor regions, autonomous energy services hold obvious value for pretty much everyone. It doesn't take much to turn commercial opportunity into functionality and form factor.

The key to all this is to sell to multiple sectors. In an energy-poor region, it could be a long time before sales become predictable to the level of confidence usually sought for high volume manufacture. (I'm talking in the millions of units.) But that misses the point: if you are versatile in development, you can overcome these concerns. Amortising development costs is not best achieved by trying to capture the greatest share of the solar lighting market alone; but by competing in the lighting and the phone charging markets, tapping into camping sector revenue, outdoor pursuits expenditure, and some of the music festival or beachgoer spending. And capturing this revenue all from the same core building blocks.

Modest volumes in the industrialised world yield
strong margins. They support the relatively low margins
but high volumes encountered in the developing world

Commercial momentum

We have seen how versatility in function and form factor can be utilised to provide significant commercial value. This can encompass profitable end-of-life recycling, addressing multiple sectors from a single core product, being able to bridge global markets and responding to unforeseen opportunities. Once we are able to look at the whole autonomous energy sector from a position of high quality and high versatility (while still being cost competitive), we can develop confidence in the truly enormous scale of the commercial possibility.

Throughout the world, there are hundreds of autonomous energy service products and solutions, offering strong value propositions for diverse types of customer. The money for these energy services can originate from an even broader range of sources, and is justified against both generic value arguments and additional value for a given sector, environment, application or user group. In fact, there are so many ways to categorise opportunity that it becomes daunting to make sense of them all. This is a real problem; for some, it is possibly the greatest problem to market entry. Multinational organisations – those most capable of supplying energy services at this scale – are perfectly aware of the potential for generating revenue in the developing world. But they perceive a lack of market entry and commercial development options to ensure continuous growth.

Here again, a flexible strategy and versatile hardware are your friends. If you can recognise the overall scale of a commercial opportunity and justify

the need to act boldly, you can establish sufficient incentive for capable organisations to enter the autonomous energy sector, or invest in smaller organisations. If you accept that success is based on capital investment to achieve
versatility in the entire energy services sector, then the challenge is no longer
about finding revenue to cover those significant development costs. We know
our versatility underwrites our ability to draw revenue from multiple sources.
Instead, the challenge is to categorise the market: to map the commercial
landscape, figuring out the business of supplying energy services – the generic
steps of growth – and plotting a commercial path against the uncertainties
and risks of operating in vast, changeable and dynamic regions.

Recall that I stated the basic needs for large multinational suppliers: strong
revenue potential; confidence that it is possible, both technically and logistically; an understanding of risk; and market growth potential. I then concentrated on growth, proposing that if potential revenues are sufficiently high,
all other barriers can be overcome. Revenue and growth are not defined by
the present market sales of solar lights or diesel for generators; rather, they
are defined by the breadth of autonomous energy service solutions which are
likely to be purchased if the value, once recognised, begins to accumulate.

By recognising the overall scale of the commercial opportunity and
by maximising its potential, you can establish sufficient incentive for
capable organisations to enter the autonomous energy sector

You could put it another way. Commercial momentum is not just about sales
and revenue, it is about brand and knowledge sharing. Every person who is
taught the value of autonomous energy services is a new potential customer.
The organisations they work for become potential customers. Their governments become potential customers. As their understanding of value, reliability
and potential propagates, so too can commercial growth. This is commercial
momentum.

If one adopts the central philosophy of being as versatile as possible – in
hardware, supporting information, as an organisation and as a brand – then
the whole world is the potential commercial market. Every person.

CHAPTER 20
Mapping global markets

Abstract

In this chapter, we consider the global market for 'personalised energy' as just four sectors: the industrialised world; disaster and emergency response; small island states; and the developing world. For each, we describe the challenges and the value propositions for stand-alone energy solutions.

The monetary value of these markets is not easily quantified. Individually, none has proven sufficiently strong so far. View them together, however, and there is overwhelming opportunity.

We also look at the collective spending across these regions by non-profit organisations on energy services.

The key realisation here is that we must consider the biggest possible picture in order to fully appreciate the opportunities – and to plan for them. It means taking the whole world into consideration, defining where generic value *lies in each sector, and between sectors.*

Scale matters. To break out of the constraints of low-cost/poor quality competition requires a broad reach, broad appeal and broad revenue potential. And it calls for a large organisation to invest serious capital to instigate the process of change.

The developing world is easily the largest market for autonomous energy services, and the most difficult to make money in.

Mapping the commercial landscape in global terms before even entering the energy services sector may sound somewhat naïve in this light. But as the tent example we saw in the last chapter demonstrates, there are compelling reasons to understand value propositions across the entire commercial landscape – in particular, how well-designed products for one sector can be applied to another. There is simply no need to aim a commercial business *solely* at energy-poor regions, when we have versatile hardware and a dynamic commercial capability. This naturally creates value for other market sectors. By including industrialised world markets in our investigations, we can more easily inspire confidence among investors. Additional market sectors could become the key to establishing a brand; and, while not commercially necessary, could provide the greatest benefit to disadvantaged people.

By including industrialised world markets, we can more easily inspire confidence with investors. Additional market sectors could become the key to establishing a brand

---

http://dx.doi.org/10.3362/9781788530705.020

The industrialised world

As we know, nations with widespread utility electricity coverage hold little in the way of value propositions for PV-enabled autonomous energy services. Most buildings effectively have an unconstrained supply of electricity. Electrical appliances serve our every wish. Away from the mains, we can charge our portable devices in our cars, or by using personal battery packs. For the vast majority of people, autonomous energy services are simply not needed.

For the decision makers who mostly live in these regions, autonomous energy holds little meaning. Consequently, the enormous and diverse value of these services is going to be underappreciated from the get-go. The energy service infrastructure we see in industrialised nations does not translate to other sectors, but this does not stop the industrialised world forcing this model on others. At present, the products out there – think of the phone chargers, PV bags, battery chargers, garden lights – are gimmicks, with consumer appeal instead of fundamental value. Limited applications are to be found in outdoor pursuits and leisure activities. Campers and beachgoers may wish for lights, chargers, fans, fridges; but these are quirky lifestyle accessories rather than necessities, and face being used a few times before being retired to the darkness at the back of the cupboard.

The most important commercial strength of an energy
services supplier is the ability to sell services universally

---

Many consider sales to the industrialised world essential for global success; though as I've already mentioned, I feel that this is more down to stakeholder confidence and perceived company value. For me, the most important commercial strength of an energy services supplier is the ability to sell services *universally* – not just to energy-poor regions. But in the meantime, we feel we do not need autonomous services in industrialised markets; and it is critical to appreciate why. It is not because they are ubiquitous. We remain unmoved by the idea of off-grid energy because, fundamentally, the utility grids are highly dependable. When this reliability starts to falter (and, given what we saw in Part One regarding the economics of grids, it *is* 'when', not if), the alternatives at our disposal will become more valuable than in any of the other sectors. Commercial development across the board is therefore essential preparation for the near future, a time when industrialised nations start to wake up to energy services which are not dependent on fossil fuels, large grid networks, prevailing monopolies or continually increasing costs.

Disaster and emergency response

The PV-enabled tent application we saw in the last chapter holds commercial value for both the industrialised world and the developing world. Arguably it is most valuable for DER (disaster and emergency response) applications.

The primary causes of large-scale disasters or emergencies are flooding, earthquakes, volcanoes, landslides, wildfires, tsunamis and, most devastating of all, warfare. All of these displace large numbers of people from their homes. A critical rapid response is to provide temporary shelter, first in large public buildings such as sports stadiums and school halls (if this is an option), or, more commonly, in tents or other small structures. (After the initial media coverage of an emergency, interest wanes and financial support dries up. The 'temporary solutions' – an oxymoron – can often be the only option for many years.) Tents can vary, from basic two-person affairs with minimal environmental resilience, to more versatile structures designed to act as short-term homes. Some are inflatable; some require substantial strength to resist wind and storm forces; and some need to withstand extreme heat or cold.

Such displacement tends to concentrate large numbers of people into cramped living conditions. This is a perfect breeding ground for disease. Clean water, refrigerated medical vaccines, sterile medical equipment, adequate lighting and cooling are needed at the earliest opportunity. The energy services required for these responses need to be rugged, water resistant, universally easy to use, supported with visual instructions and, most importantly, ready to be despatched at a moment's notice. The DER market is an important one to acknowledge for this very reason. The quicker the equipment is supplied, and the better it is at meeting needs, the greater the chance of mitigating the worst effects.

My limited experience in this area is with communications – the ability of refugees of war and famine to communicate with the outside world, particularly their families and unadulterated news. In no small way, communications protect against the effects of propaganda, the likes of which led in part to the Rwanda genocide in 1994. In a more recent example, in August 2014 the British Government sent PV mobile phone chargers to Iraq, to enable communication by isolated communities threatened by militant forces.

For all of the above, PV fits the bill. It has huge value for DER because it is inherently reliable and durable, with no moving parts, and can be used anywhere.

What are the downsides? I have met with a number of disaster response organisations in order to understand some of the demands placed on energy products in such situations. Areas heavily affected by snow coverage or the aftermath of volcanic eruptions place severe restrictions on the early effectiveness of PV. Volcanoes are particularly challenging because they throw huge amounts of ash and debris high into the atmosphere. The resultant volcanic ash can travel hundreds of kilometres, sticking around for many months after an eruption. Aside from the horrors of poisoning the air, water supply, crops and livestock, the reduction in available sunlight will disrupt solar generation, even if the panels are brought in and cleaned at night. But this ignores its equally important role as a storage medium. If the battery in the PV service is charged when delivered, it is every bit as good as a disposable battery option (the only other practical solution in the circumstances) in the first instance.

The PV services will then be usable the instant the skies begin to clear. Even against these strong odds, I am confident that PV can still meet the needs of early-response teams.

Energy services need to be rugged, water resistant, universally
easy to use, supported with visual instructions and ready
to be despatched at a moment's notice. PV fits the bill

The DER market, by its very nature, is unpredictable in terms of demand. Ethically speaking, it should have access to the most appropriate energy services possible. Nowhere else is high quality and versatility so valuable. However, commercially it is unattractive. Organisations generally do not invest the significant finance and effort into designing better energy services for a market with unknown demand. Consider that the key requirements of a PV system for emergency shelters are light, mobile phone charging and radio. These are functions which *apply to other sectors*, meaning that the development can be financed from them.

But again, this cross-sector value is only present if the products are sufficiently rugged, technically advanced (able to sit in a warehouse for several years and be functional the moment they are taken out of the box), intuitive to use and supported with materials that are universally meaningful. Millions of tents are provided for refugees each year. In addition, there are medical tents, communications tents, staff tents, tents for isolation, for food preparation, for sanitation purposes and a whole lot more. The DER sector has significant commercial potential; but the nature of this potential does not secure development expenditure, despite the ultimate human and monetary value it represents.

But, what about joining up with the potential of other markets, under the mindset of a global commercial plan? Then, the DER market can be served, and served well, without the developmental capital uncertainty. This principle is reinforced when we look at the third of our four global sectors, below.

Small Island States

Small, geographically isolated nations and populations, or Small Island States (SISs) comprise fully 5% of the world's human population. That is 300 million people – 300 million potential customers who presently suffer high fuel costs, restricted energy services and poor operational support due to their remoteness from large landmass resources. Within this demographic are Small Island Developing States (SIDS), coastal nations that experience acute challenges of sustainable development. With small populations but sizeable administrative needs, their progress is severely hampered by the high cost of energy, communications and transportation. They are vulnerable to both volatile external factors (being dependent on international trade) and to environmental disasters. They face additional threats from a combination of economic upheaval and climate change, and development opportunity is limited as a result. Small Island Developing States make up approximately 66 million people (United Nations, 2014) – nearly 1% of the global population.

The requirements of the DER generic sector (such as long shelf life and rapid deployment to diverse, unknown environments) are valuable to SISs also, and they represent positive global value. Similarly, when planning for sales to the global energy sector as a whole, SISs define certain aspects of product functionality. But I find the SIS sector especially powerful because it demands a strong understanding of logistical challenges. And it demonstrates the value of reliability, or predictability, in energy services.

SISs are, by definition, isolated from continental markets by water. However, the challenges are very much akin to those faced by remote land-based communities in the developing world. Energy is essential to quality of life and to economic progress; fuel is far more expensive and logistically trickier to manage than in high consumption regions; and even the price variations of that fuel can cause economic disruption. Skilled services for managing and maintaining energy systems are expensive and very limited.

When we think of these communities isolated by the ocean, the value of autonomous energy services becomes pretty much self-evident; more so than the places most people categorise as 'off-grid' regions. Two issues stand out, which are made easier to appreciate by understanding the SIS demographic: the logistical struggle; and the isolation in which the energy service solutions will be used.

First, the problems of logistics are easy to appreciate: everything must be delivered either by boat or by air. Then, once at the recipient end, goods are mostly manipulated by hand, but even when automation or machinery is employed, the goods experience far more knocks and bumps than they would when packed strategically within a shipping container. Goods are exposed to the elements – extremes of temperature and humidity, along with rain and snow. The roads are rarely smooth; the small, local land transportation does not have finely tuned suspension; and pallet trucks or other mechanical aids are frequently not available at the final destination.

Anything that requires maintenance is obviously going to be expensive to run. But specialist parts and the ability to fit them are often beyond the local capabilities – especially for electronic equipment. By contrast, PV-enabled systems, which require minimal unskilled maintenance, hold a value which is obvious. They also serve the need for longevity and versatility, particularly the ability for the user to define specific value from their energy system.

Unfortunately, much like the DER sector, Small Island States do not represent a primary market for those organisations currently capable of providing quality energy services. But if you combine the two sectors (even though they may not seem directly related to our other global sectors), you may see that the scale of the commercial potential grows and becomes easier to fathom.

The developing world market

So to our fourth and final generalised global sector. To clarify context, I use the term 'developing world' specifically to mean areas within large land masses without nationwide reliable utility electricity, and large populations without access, financially or otherwise, to basic needs such as education, healthcare, communications infrastructure or business opportunities. These areas do not have the defining characteristics of the DER and SIS sectors; although they will inevitably share certain common features.

For energy services, the developing world, DER and SIS sectors share many requirements. From a commercial perspective, if all three are considered at the development stage – for both versatile hardware and a commercially dynamic business model – all will produce profit, and all will benefit. Specifically, it pays to be active in the sector with the largest number of potential customers; namely, 'off-grid' locations, where there is no reasonable access to reliable energy services. Off-grid is the largest customer base in terms of numbers, and represents those disenfranchised people for whom modern energy services are most valuable. It is also, by far, the most difficult sector to penetrate and make a difference. The case for an *initial* commercial focus on off-grid inhabitants is weak; but despite there being *relatively* little money available in this group, its sheer size – at least 1.3 billion people – makes it commercially naïve to ignore.

Like the DER sector, off-grid regional markets are unpredictable, but very large – and with enormous existing expenditure. Like the SISs, off-grid is extremely taxing in terms of logistics and the lack of skilled support. Taken together, all three markets represent significant commercial value for our strategy, on account of the versatility and reliability we can bring.

It pays to be active in the sector with the largest number
of potential customers; namely, 'off-grid' locations

As the strongest value proposition, off-grid is the driving force for versatile technical solutions, the reason for embarking on commercial mechanisms of supply and a commercially dynamic capability. For our purposes, it is the target demographic with the greatest global expenditure – the demographic from which autonomous modern energy service solutions can gain the greatest share. As such, it's the reason behind this book. Unfortunately, when it comes to the developing world, an analysis of the off-grid sector reveals, not a litany of success stories, but a stout lesson in what *not* to do. So let us stop punching the air for a second; and explore why.

What not to do in the developing world

At present, the majority of autonomous energy products and services are targeted at off-grid regions – and usually at relatively poor people. By definition, there is very little spending capacity in this sector – one consequence being the lowest-cost-wins market. Products are extremely cost sensitive so again, by definition, they are of relatively low quality, low value and, most importantly, of low versatility.

I have made reference to the dysfunctional nature of the developing world energy sector throughout this text. It should therefore be no surprise that when mapping commercial sales and growth – mapping the commercial landscape of the developing world – the emergent theme is about minimising risk, the ability to overcome challenges and recognising where caution is essential.

To recap, I talked earlier about the hazards to commercial trade in these regions: the procurement specifications which guarantee over-priced, low value failures; the trader who ruins a supply relationship for a 5% saving on capital, at the expense of the project and its legacy; governments who exploit local specialists for free consultancy while awarding contracts elsewhere. In addition, there are the illogical hurdles: import duties being demanded at national borders of countries which officially do not charge on clean energy products; and non-profit 'ethical' organisations who screw over a local supplier for the sake of a few percentage points.

People have ways of defining the value of money and technology. By this I mean that people have ingrained attitudes towards transactions – preconceived expectations surrounding the supply of energy services. It is useful to understand, or at least to be aware of, some of these attitudes regarding both commercial and (supposedly) non-profit energy services provision – those who validate their actions under the far-too common mantra 'we are saving the world so anything is justified'. Similarly, many within the developing world supply sector run their businesses on a central principle: you do not need to generate demand; you need to convert it into revenue. There is so much existing spending by individuals, organisations and governments that a business strategy 'simply' needs to access that spending, rather than generate sales through new, innovative means. (A prime example of this is the initiative to replace kerosene lamps with solar lights.) Unfortunately, this approach to business has resulted in a predictably dysfunctional developing world sector. The higher the concentration of funds, the more attractive the possibility of accessing them.

It's not hard to see the appeal. If you aim to sell a single product to each of a thousand rural off-grid people, it will be expensive, time consuming and logistically challenging. Much nicer to sell one thousand products to a single organisation (over lunch if possible). These are the 'ambulance chasers', the unscrupulous suppliers who will go to great 'efforts' to win these types of contracts; anything from straightforward bribes to blatant deception of customers, who may include non-profit and philanthropic organisations. A disturbing example is the number of pay-as-you-go and tariff products specifically targeted at microfinance organisations (who, by definition, are dealing with those for whom money is most precious). Sometimes these are from deceitful suppliers who will say anything to make the sale; others are unaware of the limitations of the products they are promoting.

Academics are a particular problem regarding microfinance because they have such synergy with the finance originators. Unfortunately, neither show a strong enough understanding of the hardware they are mutually promoting. One example product of the last few years included cables and plastic mouldings which are not UV resistant – a fundamental requirement for anything designed for use in direct sunlight (a clue here: it's a *solar* product). UV stability carries negligible cost impact, but failing to address it cripples the usable life of the product. Quality control procedures and product testing would easily have revealed such a limitation, whereas an analysis of the supply cost and payback figures would obviously miss it altogether.

Unsurprisingly, the current market for off-grid energy expenditure largely follows wherever the funding is coming from. The major funders and influencers (particularly the World Bank, and national organisations such as USAID, the UK's DFID and so on) define the regions of concentrated sales, and the applications. The procurement processes which formalise the larger funding opportunities are shockingly wasteful, and they illustrate the problems with the sector as a whole. These various concentrated sources of funding have the direct consequence of *reducing* quality and value. In most cases, there are standards to which 'approved' or 'qualified' products must comply. But these are based on relatively low levels of functionality – and defined to a large extent by the price point that research has deemed necessary for the customer base. In addition, anywhere a market is being promoted, be it through 'user awareness', 'social development' or other such funded campaigns, it attracts cheaper, non-qualified and non-approved products into the bargain.

While it might seem logical to approach the off-grid market by analysing existing spending as a means of generating business, my experience and the legacy of failed suppliers show that it does not lead to success. Remember, compared with the scale of the customer base and the several decades of recognised value propositions, there *are* no successful autonomous energy suppliers. Sure, analysing where the money is coming from and looking at competitor products are important under normal circumstances. But here they do not define market potential; existing expenditure is not always about services and not even about the best power supply for the actual need. Empirically, there is nothing meaningful to analyse.

Non-profit spending

Non-profit organisation spending is immense. There are tens of thousands of organisations operating in or for the developing world. Some are very specific – for example on a women's group in a single rural village. Some are international, with a central commitment to education, healthcare or agriculture. And others are involved in a range of people and development activities. Collectively, these organisations spend billions of dollars each year on energy-related services. Some of the larger aid organisations spend tens of millions each year just on diesel for generating electricity. They raise funds from donors of all kinds – individuals, communities, sector-specific organisations, local and foreign governments, and international funds. They operate on their own, in partnerships or as part of national or international programmes. They spend these funds on individuals, specific groups, communities, services and locations.

The common theme here is that the best products don't always win contracts. Often, political mechanisms are at work – funds awarded to suppliers who don't actually make what is required, procuring it from elsewhere at a higher cost to the customer; equipment selected exclusively from the nation providing funding, flouting specific clauses in the procurement terms. Particularly troubling is where funding sourced from one religious denomination is carefully controlled to ensure 'others' do not benefit. These scenarios will be familiar enough to anyone within the supply sector. This entire category of 'providers of off-grid energy services' desperately needs reliable solutions. The non-profit, charity and aid sectors are an essential commercial consideration; not because they serve those most in need (which in a better world would be sufficient motivation), but by virtue of their expenditure. The key realisation here, for a dynamic commercial organisation with versatile assets, is that these bodies do not have to be considered as an isolated group. Nor do energy-poor regions. You will see the familiar theme evolving, but we must consider the biggest possible picture in order to fully appreciate the opportunities – and to plan for them. It means taking the whole world into consideration, defining where *generic value* lies in each sector, and between sectors.

Errors of scale

Many companies enter the off-grid sectors with ideas which look great on paper, but their plans ultimately prove far more difficult in reality. One common mistake, which may appear counterintuitive, is starting small. Most suppliers start with the small products and typically focus on small regions. The thinking goes that the market is stronger than for larger energy services – large numbers of customers, strong value arguments and achievable price points. Also, small products appear cheaper to design. However, both of these perceptions are misplaced. If we cast our minds back, we know that the portable and consumer markets suffer from intense price pressures caused by low quality products; and we've seen the resulting problem of reputation. Good products struggle to differentiate themselves from the cheap and nasty alternatives, and this severely limits their margin potential. This messy competitive environment means that the product design actually requires a great deal more investment than the basic functionality alone would otherwise dictate. Furthermore, we have also observed the high attrition rate of suppliers as they achieve a fleeting period of market leadership, before volume expansion difficulties, quality issues, distribution challenges and a lack of product development have cast them aside.

Remember that the most successful of solar lighting suppliers – of small, low-cost products – have only ever managed sales of *1%* of potential customer numbers. (To reiterate, I am not criticising suppliers, it is the market that is the problem.)

Scale matters. To break out of the constraints of low-cost/poor quality competition requires a broad reach, broad appeal and broad revenue potential. And it calls for a large organisation to invest serious capital to instigate the process of change.

Cumulative confidence, cumulative success

The heart of this discussion is about showing sufficiently large potential for autonomous energy services. Commercial profit will come from the cumulative total of a number of markets; it will not result from trying to dominate a single market. If an investment of $50 million results in a commanding share of a $1 billion market (which I believe entirely possible), then any company will be happy. Crucial to recognise is that the $1 billion does not come from a single product, application or region or customer demographic. It comprises all of these. It comes from being versatile and dynamic.

- The industrialised world exhibits its spending on 'green' and 'eco-friendly' products, but the noted absence of a value proposition undermines the business case for capital intensive product development and market growth.
- The disaster and emergency response (DER) sector has highly concentrated volume needs, and there are substantial donor-generated revenue potentials in response to a crisis. However, it is unwise to base a business strategy on an innately unpredictable sector. (Or, as humans, to be part of a venture which feeds principally on suffering.)
- Small Island States have convincing reasons to buy reliable and versatile stand-alone energy solutions, but it is expensive and time consuming to reach them. Despite strong value propositions, the logistical and brand costs cannot be justified against a small yet dispersed customer base.
- Our largest market (in terms of potential customers and spending) is the developing world. This includes off-grid regions which, for our strategy, represent the highest value propositions. But the developing world is the most dysfunctional sector with the least clarity of future revenue. Our versatility and dynamic commercial capabilities ensure revenue and growth, but we do not have the level of predictability required for the huge investments in development and pre-sales that we need. The problems do not concern technology or money. Non-profit/aid/donor spending alone is a massive source of commercial revenue, but the illogical and convoluted nature of business in these regions, along with the unpredictability of scale and locational focus, means that we cannot rely on this sector (in isolation, at any rate) any more than the others.

The monetary value of the industrialised world, Small Island States, plus the disaster and emergency response markets are not easily quantified, which is why there are so few globally recognisable companies from these sectors. Individually, they are temptingly close to offering the revenue potential and the confidence for establishing the supply of high quality autonomous energy services. But none has proven sufficiently strong so far. View them together, however, and there is overwhelming opportunity.

It is hard to see predictability in this picture. But this is also true for some of the biggest and most powerful companies of our age. Apple, Google, Amazon, Facebook, Twitter – these all grew dynamically. They have influence on their market and their subsequent revenues and growth, but they did not define their own destinies in detail in advance. They established defining principles which have been key to growth, but they were also dynamic with their products and their attitudes to opportunity, risk and overall mission.

Commercial profit will come from the cumulative total of a number of markets; it will not result from trying to dominate a single market

---

This is why visual information and brand are essential for versatility and for overall reach into the many different global markets. As a combination, they maximise exposure to potential customers, and are valuable for all applications and their users. Once developed, they significantly aid dynamic response, because they can be sent without delay to anyone – requiring no translation, no regional adjustments and at no cost. Versatile products also make you adaptable to opportunity without the significant time and cost burdens of lengthy development.

Versatility also enables growth at a predictable rate (if not over a predictable course). But the real prize comes once you have built some sort of momentum. Then you have a commanding entry position for all the commercial sectors, the multitude of applications, the discrete scales of service requirement, the different functionalities and value propositions for the diverse users, customer types, geographical regions and all the different budget sources.

Now, equipped with versatile hardware, and emboldened by a dynamic commercial capability, the next challenge is to establish a starting position and a set of principles for navigating the global market.

CHAPTER 21
Where to start

Abstract

In this chapter we discuss the practical specifics of where to start, both notionally and geographically. We outline the many ways to categorise commercial opportunity and sources of revenue within the global autonomous energy services sector.

We discuss market potential from an energy-centric perspective and illustrate that funding for energy services does not only come from offsetting existing energy spending.

In general terms, we can think of money already being spent: on 'solar'; on energy; on renewables; on infrastructure; and on OEM (original equipment manufacture). Each of these sectors represents tens of billions of dollars of annual expenditure.

Once we recognise that all electrical services are potential commercial products, we can see that our customer base is, to all intents and purposes, universal. They are individuals, households, small businesses, large companies, churches, education providers, healthcare providers and multinationals. They demand every aspect of energy services, from an individual light to a national infrastructure programme implemented by governments and commerce.

The markets, the risks, the lack of statistics

When it comes to choosing the right countries, common sense would tell us to assess risk. In Africa, at the time of writing, probably the most dysfunctional market is the Democratic Republic of the Congo (DRC). Also among those with a dubious international reputation is Nigeria, though much progress has been made in societal governance. However, avoiding these markets on this assessment alone is irresponsibly restrictive to commercial potential. The DRC has around 78 million people, but its utility and fossil fuel-derived electricity infrastructure is incredibly poor. Nigeria has a population of approximately 177 million, only around half of whom have access to electricity. Nigeria also, by the way, has one of the largest domestic film industries (who knew?) – a significant commercial opportunity for autonomous energy services.

There are many ways to research and assess markets from a national per-spective. I have chosen to work in countries which are fairly small (such as Rwanda and Malawi); relatively peaceful; and with a general lack of natural resources (which tend to foster corruption and social unrest). But this is just my personal preference. Here I want to discuss market potential from a more

http://dx.doi.org/10.3362/9781788530705.021

energy-centric perspective, particularly with respect to utility infrastructure and its reliability.

First, though, and in case there is expectation otherwise, be prepared for a distinct absence of numbers. This will be frustrating if what you want to read is where the big market opportunities are, or what products to make. As I will continue to emphasise, this is not how the developing world works. The regions are too vast and the human sources of data too diverse for the results to be representative or meaningful. I could quote sources quantifying the commercial potential of various energy service markets – be they PV lighting, phone charging, solar home systems, irrigation, Wi-Fi, mini-grids or green energy for telecoms. But to my mind, the numbers they contain are unhelpful at best; and a hindrance at worst.

A more valuable activity is to step back and look at the common themes. Every report concerned with modern energy services will give you figures in the billions, or tens of billions of potential sales; and while energy-poor regions are extremely complicated to describe in terms of business challenges, it is this consistent *magnitude* which is important to us. This strategy will appeal to those who already appreciate this difference; to those who also know how much money is out there, who understand how it can be better spent, towards more productive, environmentally sustainable, better quality and higher value alternatives.

I appreciate that this leaves little against which to measure a commercial strategy. But I do not believe it is possible to offer any consequential advice on the specifics of these generic markets, using hard numbers alone. They simply do not fit with what most of us experience in the field. Hence my conviction that any commercial venture of this kind must be flexible and dynamic. The watchword is *versatility*, which should inform the strategy at every level – from its assessment of markets to the design of its products, and beyond. This, in turn, means that significant investment is required to realise these assets to the greatest possible effect.

The grid as a starting point

The striking aspect of energy market estimates, aside from their sheer magnitude, is that they tend to *exclude* those directly connected to utility electricity. Mostly they describe direct spending on fuels or secondary mains electricity, by off-grid people and businesses.

> The initial commercial opportunity is to be found close to
> the grid infrastructure, but where dependability is absent

This is intriguing. At the outset of this book we investigated grid electricity for a specific reason: the presence of the grid defines the commercial opportunity for 'alternative' energy services. This might seem fairly obvious, but the *initial* commercial opportunity in the developing world is not furthest from the grid, as most would suppose. Instead, it is to be found geographically and physically close to the grid infrastructure, but where the true value of the grid – dependability – is absent.

The present approach to energy solutions in the developing world is to target those using fuels directly, off-grid. The primary areas for 'solar' include thermal or PV cooking to offset the use of wood or charcoal; PV-enabled lighting to replace kerosene; or larger PV systems to replace diesel generators. Logical as it looks at first, it is not actually the best way of providing long-term energy services to the developing world. Nor is it the best way of tackling fuel dependency and pollution.

The commercial targets

We can describe 'commercial opportunity' as the point where two key factors coincide, or intersect: the highest level of disposable income; and the likeliest prospect in which you can demonstrate value. Recall, if you will, our four categories of people, according to their access to electricity: reliable networked grids; unreliable networked grids; limited mains electricity access; and off-grid (no reasonable access to mains electricity). Remember also that environmentally friendly, fuel-free value propositions are only *supporting* players in our arguments. Fuel remains the standard reference point for establishing the relative value of alternative energy solutions in the developing world. (As I've established, I disagree with this benchmark, and maintain that dependability is what matters; but I digress.) Each of our four categories above represents generic fuel consumption, but they also represent relative levels of fuel awareness, motivations for changing energy services and the levels of ability to do so. What are the commercial opportunities, then, for each of these categories?

People who live with *reliable networked grids* consume the most fuel. They have the most to lose, logistically or financially, were it to be restricted. These are the people with the greatest potential to change societies. Unfortunately, they are conversely the least aware of the issues affecting the world, and are the least motivated to act. This category encompasses pretty much everybody in the industrialised world.

The remaining three categories represent the rest of the world – the developing, energy-poor, subservient nations. People who are connected to an *unreliable networked grid* suffer from fuel dependency, yet without the compensation of reliability. The more the modern world embraces electronic and online living, the more damaging a lack of reliability in their energy supply becomes to their quality of life. This sizeable group has money and they want reliability. After all, they are already paying for it. These people are already starting to embrace alternatives to mains electricity, such as PV, diesel generators, wind turbines, battery backup systems and even fuel cells. Where only *limited mains electricity access* is available, people have a direct appreciation of fuel. They are becoming ever more dependent on electricity for technology – phones, internet, social networking and so on. They are spending significant amounts of money and are motivated to find a better solution than costly and unreliable mains electricity. This group is driving sales of PV lights, batteries and 'entrepreneurial' energy solutions such as phone charging from a vehicle or commercial outlet. But the influence of this group has proven to be low in terms of political voice and commercial attraction; although this is changing fast.

Our last category, *off-grid (no reasonable access to mains electricity)*, is the one we associate the most with poverty. It includes the recipients of fundraising, but it is a demographic which cannot initiate any large-scale positive change. There *is* spending power here – it is already being wasted on kerosene, disposable batteries and the like – but it is not as powerful as the previous two groups above, and not powerful enough to attract real commercial commitment. This has proven the case for many decades.

You will notice the language of commercial power in the summaries above. This is the key to locating the first target demographic. A commercial strategy such as this requires a foothold, in a market which wields not only spending power, but the means for qualitative change, by realising the manifest *value* of energy products. In order to help off-grid people, then, it is necessary to start where the spending potential can be 'converted'; not merely into quick returns (this is not a land grab), but *invested* into building commercial and social growth; starting with a measure of connectivity and a dash of influence. In plain terms, it means starting in the middle and working outwards.

The first target

Put simply, utility grids do not reach rural or remote communities at all. Meanwhile, for urban areas, or semi-urban areas (on grid spurs), utility grids are sub-optimal. The former group have the greatest need; the latter group has the most spending power. The most accessible customers for quality energy services are those who spend the most money on, yet suffer the highest disruption from, unreliable networked grids. Commercially, this is where we must start. As I have said, there are plenty of reasons why grids will become less reliable, both in the industrialised and developing worlds, and few tangible indications to suggest that they will improve. We can be confident that our initial customer base is only going to grow.

All people with reliable grids (for example, the wealthy nations) are potential customers of the future. But so too, more immediately, are those currently unable to meet the capital cost of the quality products we supply. However, as the larger target market, products can be made more cost-effective to bring to these dispersed potential customers once sales have been instigated.

The most accessible customers for quality energy services
are those who spend the most money on, yet suffer the
highest disruption from, unreliable networked grids

Unexpected customers

So, is there really a demand for autonomous energy services by those who already have mains electricity? Indeed, and this is more than conjecture. I have sold a number of these products myself, and what surprised me was just *how much* money people were willing to spend on reliability. Unfortunately, this customer base is still not well understood or served with suitable products. Let us look at a couple of examples from real life, where the value of the product caught the supplier unawares.

As mentioned in Chapter 7, I was involved in the development and sales strategy of the Sundial PV lighting product. Much like the hundreds of similar items in the developing world, ours was designed primarily for people who did not have mains electric lighting in their homes. Conforming to the ordinances of the marketplace, our sales literature trumpeted its value against the alternatives it replaced: candles, kerosene lamps and disposable batteries. We were keen to distinguish ourselves from the cheaper competition. Our quality multi-functional product could be used as a desk lamp, a torch, a wall light and a ceiling light; with sufficient output for a typical room, and with the capacity to operate from dusk until at least 10 p.m. The light was rugged, and came supplied with a remote control (which itself incorporated a torch).

Curiously wrong-footed, we were surprised at how many households with mains electricity were purchasing this PV light. Because the mains grids in developing countries can be so unreliable, many well-to-do households were resolutely fed up with being plunged into darkness and forced to revert to the very candles and kerosene lamps their hard work had allowed them to replace. They were also far more reluctant to use naked, smoky flames, now that they enjoyed soft furnishings and clean-living environments. The PV light was hung from the ceiling of the living room and the remote control sat on the table next to those for the TV and stereo system. When the power went out and the room became pitch black, they reached for the remote control and switched on the light. Sales to on-grid customers outnumbered those to off-grid customers for some time.

In another example, a friend in the solar energy sector in Kenya was asked to supply a fairly large battery system to a house in one of the suburbs – again, a residence which already had mains electricity. The first thing he noticed was the smashed screen of a large TV with a marble ashtray on the floor in front of it. The power had evidently cut out just as the owner was watching the penalty shootout of a football match. The owner didn't care how much the PV backup system would cost him, as long as it was guaranteed to provide power – and he was very specific – for at least 3 hours. Whether this requirement covered the lifting of the trophy and the pundits' analysis is not recorded, alas.

Both of these examples illustrate that funding for energy services does not only come from offsetting existing energy spending. There is money available if the market is sufficiently well understood and the product meets the requirements. Take a leap of imagination here: in India for example, how many satellite TV systems could be sold if they were guaranteed not to cut out at any time during an 8-hour cricket test match? India has over 240 million households…

Categorising the commercial landscape

When we talk about 'spending money on energy services' there is an assumption that we mean direct sales to people. In fact, the definition is far broader. It refers to payment, daily usage and ongoing maintenance, and encompasses every form of electrical device or service in a given region, country or even continent.

Take a moment and look around your home, office, local shop or any public facility. Energy services permeate every aspect of our daily lives – our living, working and social spaces. Potentially, everything – and I mean everything – can be provided by autonomous energy solutions. For example, a row of autonomous street lights can replace those powered from the utility grid. Each home can have energy efficiency applied individually, enabled by their own on-site energy generation. This may appear some way off for high-consumption industrialised society homes, but it is already a practised option for developing world living.

Lighting, cooking, heating, environmental cooling, refrigeration, healthcare services, educational facilities, television, radio, computing, internet access, telephones, device charging. The manner in which these services are delivered may differ from the present, but they can all be delivered through autonomous means.

Now this revisits one of the touchstones of this book – that people require reliable services more than a mere energy source. And if every service without exception can be realised autonomously, our commercial confidence demolishes one critical barrier. In particular, we can abandon the old idea that, once the desire for energy reaches a certain scale, the only recourse is utility electricity. The myth that the grid is a necessity falls like a house of cards.

What does this do to define our customers? Once we recognise that all electrical services are potential commercial products, we can see that our customer base is, to all intents and purposes, universal. They are individuals, households, small businesses, large companies, churches, education providers, healthcare providers and multinationals. They demand every aspect of energy services, from an individual light to a national infrastructure programme implemented by governments and commerce.

Sources of funding

With this in mind, consider where the funding might come from. In general terms, we can think of money already being spent: on 'solar'; on energy; on renewables; on infrastructure; and on OEM (original equipment manufacture). Each of these sectors represents tens of billions of dollars of annual expenditure. (I am deliberately excluding money spent addressing climate change, poverty or the environment; critical and emotive humanitarian concerns, but impotent as commercial motivators.)

If every service can be realised autonomously, our commercial confidence demolishes the old idea that the only recourse is utility electricity. The myth that the grid is a necessity falls like a house of cards

---

Within these categories, those spending money on energy-related services include: international organisations; overseas governments; national governments; commercial organisations; NGOs; community groups; households; and individuals. Everybody within this target demographic is a purchaser of reliability and value. Recall our entry point to the market: we are targeting those who are already spending money on energy but are not receiving reliability (especially those spending on low quality grid electricity, or generators).

Earlier, I used the example of a remote control solar light (the Sundial), used in a home with unreliable utility electricity. Now, from this humble domestic application, try casting wider, and consider the many regional and national applications where reliability represents significant value. These include mobile phone networks, internet providers and medical services – particularly cooling and refrigeration. Infrastructure must be dependable to offer true value.

Meanwhile, remember that value is about more than just saving money, and it is often difficult to quantify. Consider the significance of all government offices in a country having reliable energy services. There is a value proposition solely in the guarantee that the services will be working come the next election.

An example: banking networks and synchronisation

Banking networks in the developing world are a prime example of how dependability is needed for regional infrastructure, yet the value may not be obvious. Most bank branches need energy for lighting, computers and internet access as a bare minimum (before you consider air-conditioning, printers and so on). Prompted by the success of microfinance programmes, banks know they need to have a presence in rural areas; but the utility grids do not extend this far out. Fuel-based electricity is the expensive alternative.

Here is a strong case in itself for autonomous energy services. But there is more. These rural branches do not assume a 24/7 connection. Typically, they can utilise a computer to record transactions in an offline manner. They then must synchronise with a central server at least once per day to ensure that transactions are passed to the bank. This simple procedure becomes a scheduling challenge, with hundreds of rural branches allocated a dedicated times lot to synchronise.

If the power is out at the crucial moment, either at that branch or at the central server, you cannot simply wait until the power returns and resume your place, as this will have a knock-on effect on all the other scheduled syncs from the other branches. The remote bank does not need to have power 24 hours per day, but it is of premium value to have *guaranteed* power at the specific time slot every day. The value proposition for the autonomous energy service – computer and internet access in this case – is predictability and controllability; and this principle of synchronisation applies to government offices, medical centres, schools, shop franchises and more.

Wants and needs

In this book we have, inevitably, talked of scarcity and of poverty. Similarly, the vast majority of material concerned with developing nations is focused on what people need. It is much rarer to find insights into what people *want*. A further reality to accept is that customers will spend considerable amounts of money even in the absence of an actual need.

Relatively poor people in rural areas *need* clean and safe drinking water but what they *want* is a cold beer or soda. Those with access to electricity need lighting to study after sunset, but what they want is to watch television. This is no different from privileged nations where we have access to the world's information at our fingertips, and where consumer preference indicates reality TV or sharing pictures of kittens on social media. This faintly taboo absence of a balanced discussion on wants versus needs is no doubt due to funding for projects in the developing world through donations and loans. I don't suppose a fundraising initiative to help poor Africans enjoy a beer would be particularly successful, however happy the recipients might be. In developing nations, discussions are understandably biased towards needs rather than desires.

In industrial societies, consumers are free to pursue what we want, often to the neglect of our needs. We develop complex emotional attachments to high-profile branded goods, whereas we have no emotional engagement with the essentials upon which our very lives depend. (How many coal mining companies, or mobile communications hardware firms can you name, as opposed to, say, fizzy drinks or clothing brands?)

This skewed attachment or sense of engagement is borne out by our expenditure. We actively spend enormous amounts of money on brands, but it is only indirectly (under obligation and reminder) that we enrich the back-room organisations who keep the lights on and the water running. All the while, governments have to subsidise (or at least incentivise) the infrastructure suppliers, in the national interest.

Those with access to electricity need lighting to study
after sunset, but what they want is to watch television

Back in the developing world, areas of essential urban infrastructure such as water, sanitation, grid electricity and basic communications are indeed high-profile spending areas. Compared with rural electricity, they receive a degree of international attention.

But that does not mean all spending is on needs. In what we know as 'poor countries' of course there are far fewer brands with lower levels of spending than their lavish western counterparts. But nonetheless there is a huge amount of 'luxury'[1] consumer expenditure. At its extreme, this manifests itself as a stark mirror of the affluent world: the flamboyant, largely credit-fuelled revolution in men's street fashion in Congo (recently co-opted by a Guinness ad campaign); individuals who forgo a meal in the day for the sake of a download; $100K cars driven daily through destitute slum villages; luxurious houses with huge flat screen TVs and satellite uplinks, filled with imported wines, foods and trinkets, a street away from scenes of malnutrition and suffering.

Where am I going with this? Essentially, I am making the point that money is available in these regions and countries; and this represents a vast array of opportunity. If commercial energy services can be made sufficiently attractive, available and affordable, they can benefit from money which is presently being spent on other things; and by this I don't just mean money being spent on other forms of energy.

Demand and priority

The scale of spending is already enormous, but it is not easily translated into specific energy service requirements. If we were able magically to present all the modern energy service options (torches, reading lights, home lighting, TVs, radios, refrigerators, phone charging, internet access) to all those now spending on candles, kerosene, disposable batteries and the mains, we do not know exactly where we might redirect this mountain of existing spending.

This is one of the primary reasons for making this strategy so commercially versatile. There is no overwhelming majority when it comes to demand. There may be overwhelming areas of *need* – but that is not the same thing. Also, without a doubt within the energy sector there is an overwhelming emphasis on supplying PV lights, but this is more to do with development spending and the economics of supply, rather than demand. In an ironic twist, although lighting is fundamental for development, it is known not to be the priority *demand* area. Hence the tens of billions of dollars still being spent on candles and kerosene for lighting by people who now have mobile phones.

Demand and revenue potential

Our initial customer base includes those who want, and can afford, the lifestyles common to more affluent societies. If we make services accessible, convenient and affordable, we will create a market for every labour-saving and lifestyle device that we all take for granted. Again, these go beyond individuals and their habitats; they are for offices, hotels and commercial outlets. They span the breadth of the leisure and recreation sectors.

Hopefully, the examples of the banking network and the 'needs and wants' market have helped us appreciate that we can create demand through the provision of services that are simply not possible from other energy systems. The search for revenue opportunities goes further than looking at what is already being supplied. Portable refrigeration is a prime example of this. Commercial opportunity is present throughout the world, in each of our four generic sectors, for every type of customer (individuals, organisations and governments), and for all the energy services familiar to us in the privileged world. It is ready to serve needs and wants; the necessities of commerce and the desires of comfort.

Multinationals

It is inevitable that a large manufacturer will supply PV-enabled services in the tens of millions of units into the developing world. It is just a matter of time. Some multinationals have already dipped their toes in the market: Total, the oil and gas giant; Schneider electric; Philips; and Samsung, to name just a few.

The example of Samsung is an interesting case study. The majority of their product range requires electricity, to operate or for recharging; and in 2012 they employed the high-profile UK Premier League footballer Didier Drogba (a native of Côte d'Ivoire, West Africa) as the face of their African advertising campaign. (I was amazed to see that during a 2012 stay in Nairobi, Kenya, you could barely go 200 metres without seeing his wealthy smile endorsing a Samsung product.) Samsung went as far as incorporating PV panels into their mobile phones, video cameras and notebook computers. While this design is a misfire in practical terms (one simply does not leave a phone face down in direct sunlight all day to charge), it nevertheless demonstrated an appreciation of three things: first, that there is a large market out there, justifying a multi-million dollar product development and advertising campaign; second, that there is a lack of electrical infrastructure which requires the devices to contain their own energy solution; and finally, a statement to the value of PV, as the most appropriate energy solution for a personal product sold into energy-poor markets.

But the brand will only get you so far. While Samsung and their multinational counterparts have shown by their initiative that money poses no problem (on the supplier or the buyer sides), they have had only limited sales success. Over and above a product's superficial desirability, even a small practical restriction or inconvenience becomes a critical obstacle in this more exacting market. In our example, many of these mobile phones feature crystalline silicon PV on the back, requiring reasonably strong sunlight to function efficiently.

Commercial opportunity is present throughout the world, in each sector, for every type of customer

---

Product design has to be defined by the needs of the users and the environments they live in, so a local, cultural understanding of practical requirements is essential. And where such a requirement exists, be it a feature capability, or a crucial dependency on direct sunlight, this has to be conveyed clearly and visually – without words and in a universally accessible manner. As I explained earlier, text, even if translated into a hundred languages, will still appear as meaningless marks to large groups of customers. The same is true for sharing knowledge of products, technology and opportunities. It is a tall order, but people in all locations need to be aware of what is available to them to improve their lives, in order to show a desire to buy.

A different type of challenge

All the technology exists to meet the tasks ahead; and commercially attractive markets are wanting. Meanwhile, the challenges facing the manufacturers are no more complicated than they would be in more traditional markets; they are just different.

You cannot simply produce an 'off-grid' product in the traditional way, initiating sales in urban areas with a view to propagating outwards to the rural regions. You cannot reach the majority of your customers through television or the internet (despite their growth), which makes it trickier to establish a popular brand identity in a one-stop-shop format. You cannot sell direct to your customer, either online or face-to-face, and you cannot reach them all in their own language. But customers – big and small, private and commercial, rural, urban and national – can nevertheless be reached. They can be informed and can be sold to. And they will keep coming back to buy more, larger and higher value energy services. Accordingly, they will become advocates of the brand in word-of-mouth societies. Quite how this is possible forms the next part of the book.

Endnote

1. The word 'luxury' evokes opulence, but is used here in a strictly economical context, denoting a class of items which contradict, or are an exception to, the law of supply and demand. We spend more on them because they are unnecessary and ephemeral (and can be status symbols, to boot).

PART SIX

The commercial value and need for local partners

CHAPTER 22
Selling guaranteed energy services

Abstract

This chapter discusses how to establish an attractive proposition for users: guaranteed services.

We do this by engaging with local partners. When I talk of local I mean people and institutions who will stay the course in the years and decades ahead.

The ability to guarantee *the reliability of energy services is powerful for attracting our initial customer base – those with unreliable networked grids.*

This group consists of hundreds of millions of people, who have spending capability but poor-quality electricity. Lighting, communication, TV, radio, refrigeration, computing, internet access – all of these services are seriously undermined if they are not reliable, or at least predictable.

The opportunity before us, as detailed in this chapter: to showcase the ways and pitfalls of energy-poor markets to the largest companies; to point out that carefully designed hardware and the utilisation of local resources need equal attention; and to demonstrate the brand value that comes with guaranteeing reliability.

Energy, on its own, is a raw commodity. So far, we have noted the varying ways of turning this crude resource into useful and manageable energy services. We acknowledge the dominant role that these services play in the lives of people, be they urban sophisticates unaware of the provenance of what powers their affluent lives, or rural poor families diligently working to keep destitution more than a stone's throw away. What we discover is that all people share a common need, not necessarily for the novelty of technology, but for something dependable that will genuinely benefit their lives.

Investigating the products themselves reveals better strategies to conceive, design, deploy and procure this equipment to the greater good – less an abstract, well-intentioned 'virtue', and more of a tangible, genuine metric of 'quality' and 'value' which promises – and delivers – reliability and longevity of use. Furthermore, it becomes clear that anyone considering a commercial venture of this type must be versatile in order to roll with the punches of tricky emerging markets. They must also comprehend that 'starting small' in the giant unlit continents is tantamount to no start at all; they need large-scale investment to adapt to the jaw-dropping commercial potential out there.

We accept, then, that it is the reliability of energy services that proves vital to development. But equally essential to commercial value is a form of corporate reliability; a quality that encompasses stability and longevity, and also

http://dx.doi.org/10.3362/9781788530705.022

responsibility and commitment. So what form do these qualities take, and how do we realise them?

The ability to *guarantee* the reliability of energy services is powerful for attracting our initial customer base – those with unreliable networked grids. This group consists of hundreds of millions of people, who have spending capability but poor-quality electricity. Lighting, communication, TV, radio, refrigeration, computing, internet access – all of these services are seriously undermined if they are not reliable, or at least predictable.

It is inevitable that multinational corporations will engage with the developing world. Indeed, this book has pointed out some of the reasons why this is critical to success: their ability to develop high quality consumer products; an existing global logistics infrastructure; and the means to apply vast capital outlay. This wholesale embrace of global giants may have raised some eyebrows; particularly among the traditional liberal consensus associated with development and humanitarian issues. Multinationals are not universally seen as a good thing outside of their immediate commercial context; the manner in which they produce, sell, supply or support energy services poses a significant impact on the developing world, and indeed the world economy as a whole; and this gigantic responsibility merits considerable critical scrutiny.

For our purposes, the guarantee of services expresses perhaps the central reason for entering the autonomous energy sector in the first place. And it means two things. First, asserting the longevity of our commercial commitment to developing regions by engaging in partnerships with local people. And second, using this resource responsibly to ensure the technical and mechanical performance of the products we sell.

The guarantee of services expresses perhaps the central reason for entering the autonomous energy sector in the first place

---

Manufacturers and local people

Local people are instrumental in the quest to alleviate poverty, and to promote social and economic development. Similarly, the key to successful deployment of autonomous energy services is engagement with the local population. By this I mean genuine partnerships, where there is necessity on both sides and a mutually supportive, lasting commercial motivation.

The reasons for this encompass all aspects of self-determination and improvement of life: political stability, transparency of governance, social development, economic strength, healthcare and so on. But above all, 'people on the ground' provide permanence – the promise of sustaining a solution into the future. So, when I talk of local people (and I shall, at greater length, in the coming chapters), I am not describing the usual approach of hiring some locals to act as sterile agents or delivery firms. Nor do I mean companies from overseas who will depart if the commercial market becomes difficult, or individuals who are not likely to stay in the long term. I mean individuals, locally established companies and governments: people and institutions who will stay the course in the years and decades ahead.

The underlying objective of this is to ensure any success benefits the people and the country. The core motivator, of course, is financial profit. I therefore want to highlight the commercial value local people represent to multinational suppliers – in short, how they enable profit. More importantly, I want to show the immediate and lasting financial reasons why multinationals need to engage with locals of all kinds as partners, not as passive task providers.

The single obvious justification for employing local people in a vast new market is, simply, that they are local – nearby, numerous and on hand. With a need to draw revenue from multiple applications, sectors and regions, it is unrealistic across the distances of the developing world for a multinational company to supply people into every geographical location in which they wish to make sales. When it comes to getting established in developing regions – legally, commercially and as a brand – the initial need for utilising a regional population may seem clear enough. Still, commercial companies have a tendency to try to reduce their external costs once their businesses become established; and in the interests of efficiency, they often employ lower quality methods of supplying and supporting their products.

The temptation to reduce costs is great indeed: abandoning the expansion of physical shops in favour of online support; digital media instructions with limited language support; centrally based collections and returns facilities instead of a network reflecting customer distribution; reducing the range of products and solutions (such as battery or PV size options); leaving product setup to the user; and a range of optional accessories, not supplied as standard. This way, large suppliers can 'control' their operational costs in the face of growing competition and the continuing price concerns of the market.

This, I believe, is false logic – short-term savings for long-term losses. Remember that this strategy is conceived as a whole, and will only achieve a momentum of commercial profit and continued growth if it is implemented also as a whole, with a view to long-term outcomes. This is easier said than done. The principles of this strategy require significantly higher investment upfront; it is more expensive than simply trying to grow the business organically from one or two products. Hopefully, I have justified the value of versatile product development, visual media and brand against the ability to draw revenue from multiple markets, applications and budgets. Now I want to describe why yet another additional upfront cost is justified – that of motivating local people financially from the very start.

There are two functional aspects to the profitable sale of quality autonomous energy services across the globe. First, we have to develop and manufacture the appropriate products and solutions. Dozens of commercial multinational manufacturers (or organisations who can function like multinationals) are entirely capable of producing the product ranges. Second, our service needs disseminating in the form of sales, support and growth, for each geographical, social and application region. This is more complicated in the developing world than in industrialised societies, and especially compared with more standard 'supply' arrangements. For a start, the challenges of engaging with millions of people within each sector, within such a diverse target customer base, are considerable indeed.

This strategy is conceived as a whole, and will only achieve commercial momentum if it is implemented also as a whole, with a view to long-term outcomes

---

This is not all: in turn, we need each of these local representatives to be 'on message', in order to communicate the value of quality energy services, and to help overcome those generic barriers of the market – those assumptions, misconceptions and cost obsessions that we have discussed. Thankfully, this will have been made easier if we have done our homework beforehand. If we have developed versatile, quality products, whose value is made manifest by the use of visual interfaces and data collection (and supported by media conforming to our Visual Instruction System), then it becomes entirely feasible for local people to supply and support energy service solutions and create commercial value for the original manufacturer.

But this comes with conditions attached. As ever, the problem is with people: those of us who 'have' defining the roles of those who 'have not'. Put more plainly, it is essential to lasting growth that we see and treat local people as *partners*. Not delivery networks, agents, sales outlets, support centres or mere guides when head-office personnel grace local areas with a visit; but as fellow agents of commercial value. We will see the importance of this later.

Reliability and value

The 'people factor' does not end here. People determine more than just a company's stability; in fact, they determine the very reliability of the products themselves. How so?

The user must engage with a PV-enabled device to understand and obtain its true value. This is one of the principles of 'Personalised Energy'. This is the same as saying that a device needs a person to interact with it correctly in order to ensure its reliability. So, we have to support customers in their understanding of PV-enabled services. No matter how good our visual media is, there is always value in having an actual person to relate product and media to individual circumstances. With basic knowledge, an individual could fix common problems, and feed information back to the original supplier, thus creating value in the eyes of the supplier.

I have witnessed this exact scenario. In a rural high-altitude village in Rwanda, we discovered that PV modules were failing in reasonably high numbers. We had heard that the product owners were treating their PV panel with great care; in most cases wrapping it in the original packing material every night. But during a visit we were dumbstruck to see the villagers approaching us swinging the wrapped PV modules by the cable. The PV panels looked spotless, but the output wires from the PV were failing due to mechanical stress. It was then a simple job to explain, via an interpreter, how *not* to carry the PV; and an enthusiastic local boy subsequently made sure other owners knew how to handle the equipment. There were no further problems of this kind.

Scenarios such as this were overlooked by the module manufacturers; over 90% of PV installations are grid-connected, where the module is installed and connected, and the cable doesn't move for the life of the system. Rigorous design and intuitive visual media can address these kinds of issues, but the individual in the field will still be needed for the next unforeseen problem.

On its own, this after-sales support is not a strong enough commercial justification to put people in the field. But remember that we are deliberately committing to selling high quality energy services into generic markets obsessed with the lowest cost option. Justifying the extra cost means that we require people to convey to potential purchasers exactly what this quality means; and how engaging with a well-made product will *guarantee* its performance and reliability.

People determine more than just a company's stability; in fact, they determine the very reliability of the products themselves

---

Making batteries last longer

The ability to guarantee PV system performance is rooted in the techniques for making the battery last longer. In basic terms, increasing battery life requires using less of the available battery capacity on a daily basis. (To clarify, here I am talking specifically about larger systems, in which the battery is designed to deliver service throughout the year, in all weathers, and for systems where the battery is usually stationary. Personal and consumer products, meanwhile, are *entirely* dependent on user understanding and operation; although local advice, maintenance and repair are of course valuable.)

Extending the life of a lead-acid battery[1] is achieved by reducing the amount of energy drawn for each charge-discharge cycle. For a typical stand-alone PV system, charging occurs during the day and discharge occurs at night (or also at night). So, one cycle is one day.

There is a direct relationship between the daily depth of discharge (DDOD) and the cycle life. A typical PV battery might offer 800 cycles (just over 2 years) when discharged to 50% each day, but 3,000 cycles (just over 8 years) when the DDOD is less than 20%. The consequence of using a smaller percentage of battery capacity is that the battery needs to be bigger. If the daily load requirement of the energy service being provided is 200 watts, the battery is 400 Whrs when DDOD is 50%, but 1,000 Whrs if DDOD is 20%. This of course means the battery is more expensive – another reason why the energy services of this strategy will likely not be the cheapest.

In addition to increasing battery cycle life, the lower the DDOD, the greater the battery's reserve energy level. This is known as battery autonomy – the ability of the battery to continue to support the load equipment without receiving charge from the PV (or anything else). Battery autonomy is important because it enables the load equipment to be supplied overnight, during periods of bad weather and throughout the seasons, when the daylight energy is lower than needed. In high latitude locations (such as Northern Europe and Canada) battery autonomy can be over 30 days – meaning the DDOD can be just a few per cent. This is to ensure that the battery can support the load throughout the winter, when it will not receive an adequate charge from the PV for extended durations.

Industrial off-grid PV systems are so reliable largely because of their careful battery design, which accounts for annual climate conditions and the required life of the battery. This reliability often becomes compromised in the developing world; the preoccupation with cost causes pressure to reduce the size of the battery against the daily load energy requirement. But this need not be the case. Indeed, long battery life and autonomy can be exploited as a sales feature.

Earlier, we looked at microprocessor design and its benefits for product development. The core microprocessor approach enables detection and warning of a number of possible faults in the system. For example, a simple light sensor can be used as a reference against PV generation. If it is a bright day but there is no or lower than expected generated energy, there is a problem on the PV side of the system. If energy is being drawn at an abnormal rate, given the known devices connected to the system (which can be detected or pre-set into controller memory), then this calls for analysis on the battery output or user device side of the system.

Our microprocessor can notify of the possible fault via the user display, an audible alarm, a text message, an email or to a central internet database, subject to communications capabilities. (Some of the older industrial systems used to have fax capabilities.) If a fault condition is signalled and the system has been designed correctly – with several days' autonomy, there will be around three days within which someone can visit and address the condition. Not enough if a small support staff is based in the large cities (as is common among energy suppliers); but ample time for a local professional to reach the installation. In short, we have a pre-failure warning system.

Pre-failure notification is not a theory; it has been practised for decades; this is how industrialised off-grid PV cemented its value in the 1980s

---

The most common faults in stationary stand-alone PV systems are either dirty or damaged PV modules, or excessive energy drain. Both of these conditions are especially common for solar home systems where the PV module is brought in every night for security reasons. A modest understanding of the service allows a local partner to visit the system; either fixing it on the spot, or making an informed inspection (including extracting data from the controller if necessary) so that replacement parts can be obtained or a more skilled local person mobilised.

The ability to apply support before any loss of service occurs is of immense appeal to customers – particularly wealthy or commercial ones. While realistic expectations must be managed in this area, the fact remains that pre-failure notification is not a theory; it has been practised for decades; as it goes, this is how industrialised off-grid PV cemented its value in the 1980s. The value lay in having the time to react when the service was safety-critical or, more likely, to avert expensive interruptions.

Hardware and support

The approach being championed here is that it is not just hardware being sold; it is a service which entails the twin elements of hardware and support. Customers need to know the value of good design and quality energy services, but also that support is essential for all products. Support can come from an individual in the village, a kiosk in a nearby town, a shop in a semi-urban centre or a city-based company. The supplier and support network are part of the same value offering to the client, and the message must be very clear: the customer is buying a reliable service, not a power supply. That service includes the user appliances and everything else required to meet the stated need.

This is how multinationals become established. Their brand can represent something that no other is able to provide. This does not have to be a financial burden. On the contrary, by truly embracing the principle of local partnership, the resulting brand value provides a potential financial saving for the original hardware; savings which can offset the costs of the local resource itself.

A supplier can decide this balance between system reliability and support during the design phase. Recall the example I described earlier, where 20 computers for a school, quoted at around $40,000 when computers and power supply were procured separately, could cost less than $10,000 when considered as a holistic solution. That $10,000 includes sufficient margin for the supplier and the estimated support requirement. If that school is 20 minutes' walk from the local partner's office, then regular PV panel cleaning, calibration of the user display and even battery replacement can be performed regularly and easily. This can reduce the cost of the hardware by 20% or more. But if we imagine that school to be 3 hours' drive, followed by, say, 40 minutes' walk up a mountain, the system would be designed for high reliability, long life and local members of the community are trained to perform the routine maintenance. To facilitate guaranteed reliability a GSM monitoring system is incorporated for remote monitoring by the supplier. The system batteries can be designed to support the user equipment for over 5 days without charge so the potential for pre-failure corrective action is strong.

Hardware is cheaper when local support is easier and more frequent, and more expensive if support is restricted to annual visits or fault condition response. This is exactly how industrial PV systems have been designed and supported for the last three decades or more.

This model works for most applications. The supplier partnership (manu-facturer and local) makes a profit by supplying a service with a stated level of reliability. When designed correctly, system maintenance is low-cost, minimal retro-active expense. The support network is economically retained for a given project, and in addition to the money they make from sales, they feed ideas and opportunities back through the network to aid growth.

The supplier and support network are part of the same value offering to the client; the customer is buying a reliable service, not a power supply

---

Sitting in the comfortable industrialised world with our always-there, always reliable, don't-need-to-think-about-it dependability of energy services, it is easy to forget what reliability actually is. But, against a backdrop of regular power cuts and subsequent disruption to every aspect of life, its meaning – and value – become immediately apparent.

The incentive of product performance

We have seen the impact of low-quality or unreliable energy on users. It also negatively impacts companies. Multinationals are driven to make commercial profit and sell new features and versions of products to the masses. Perhaps their strongest motivation to provide dependable energy services is avoiding a bad reputation. Unreliability, wherever it manifests (and whoever causes it) will be most commonly associated with their brand, undermining its value. This is a powerful incentive to pursue quality in energy-poor regions. More importantly, it can help multinationals make the initial commitment to the local energy services market. And with it, hopefully, a commitment to the principles of this strategy.

Large suppliers of electrical equipment are well aware of the energy-poor world – people without electricity, unable to use their products; the appar-ent absence of a market. The challenge is to state authoritatively that this is not so; to convince these organisations of the value of autonomous energy solutions (so they don't just fit a utility electricity plug on the end of their equipment and assume that is useful to everyone in the first instance). We need to extol the value of PV-enabled and personalised autonomous energy services. One way of doing this is to highlight some of the practical problems they are already facing.

Fictional hours of operation

Our smartphones, tablets, notebooks and laptops are all defined by their battery life, which reduces at every stage of ownership. A smartphone can have 'up to' 20 hours of operation – but only if you barely use it. Switch on cellular data, and watch lots of video and you can be down to just a couple of hours of battery service. As the months progress the battery performance only deteriorates. Charged well each time and a few years of good performance are possible. Charged badly (say, to under 80% each cycle, while being used) and the realistic battery life expectancy falls to more like 2 years.

For anyone who doesn't spend their waking lives upgrading their device, this is inconvenient in itself, despite most of us having mains electricity, a vehicle charging point or a battery backup product. In regions without mains electricity for charging, 'inconvenient' is an inadequate term. Typically there are few vehicles and likely no decent battery backup products. To make matters worse, the battery usage times are typically shorter in poor regions, thanks to the temperature and the distances from the internet or cell phone transmitters.

Batteries and electronics in general don't like extreme temperatures, such as those found in the majority of energy-poor regions. Most personal appliances today use lithium batteries. If you look at the warranty conditions on a typical phone, the device is likely to have a 24-month warranty, but the removable battery maybe no more than 6 months. And the conditions of the warranty state the device should only be used in temperatures between 0°C and +40°C. In sub-Saharan Africa, the average temperatures can reach over 40°C and in direct sunlight a device can reach over 65°C. (I regularly measure PV panel and other equipment temperatures.) Battery warranties are therefore void for many users, even those with recourse to a warranty. More importantly performance drops significantly with continued use in these conditions. The consequence is that the usable battery capacity could be reduced by as much as half; which, by definition, will be consumed more quickly.

Phones and cell towers

Let us continue with mobile phones by way of example, to demonstrate modern devices in the energy-poor areas of the world. Our phones communicate with local 'cell towers'.[2] These physical towers typically have several transmitter-receivers, each providing coverage to a given area, or sector.

These cell towers represent the front-end of our provider's overall network. The signal strength bars on our devices refer to the strength of the signal with the nearest cell tower sector, with which we are registered at a given time. Cleverly, when we move from the range of one cell into an adjacent one, the network hands over gracefully from one cell to the next, without notice to the user, even without interruption should you be in the middle of a phone call. (You used to be able to hear this electronic 'chatter' if you left your phone against a speaker or radio – the phone announcing itself to the network, and the nearest cell acknowledging your number.) When somebody sends you a standard SMS text message – still the most common form of mobile communication in the developing world – your service provider channels it to your nearest cell. When this in turn transmits the data to your device, it also asks for confirmation, a system-level 'Have you received the text?'. Your phone replies 'Yes, got it', a process known as 'handshaking'. Under normal operation, that marks the end of the transaction. This is all very nice, but in rural areas this 'handshaking' is often a more troubled affair, when a cell tower is likely to be many kilometres away and the phone lacks the power to connect. (This is made worse in bad weather or prohibitive terrain.) In the absence of a receipt confirmation, the tower keeps sending and the phone keeps responding with a signal of insufficient strength, draining the phone battery all the while.

What also happens, and why your phone battery dies more quickly in poor signal areas, is that if your phone only has one or two bars of signal strength, it boosts its transmit power to try to compensate – using more battery power. This applies to calls as well as texts.

This is no longer a 'first world problem', but is widespread in the regions which concern us here. On smartphones – the devices most coveted across the globe – texting is managed by an app on the phone instead of the operating system, but the same handshaking problems occur. But for smartphones the greatest use of power is the screen and processor. Consequently, communicating using something like Facebook will use significantly more energy than an old-style phone with a low energy LCD screen. More modern messaging services, such as WhatsApp, Facebook Messenger or Apple Messaging are even worse for battery life. They work via an always-on data connection to the nearest cell (unless there is an accessible Wi-Fi signal present).

The use of more advanced features adds to the problems. Newer multi-media-enabling technologies such as 3G and 4G use higher frequencies to allow greater data bandwidth to the device, but these higher frequencies do not travel as far as a 2G signal. Some 2G cell coverage areas could be 35 km in diameter, but a 3G cell could be quarter of this, or less still, to achieve the required data bandwidth. This basically means that operating in full 3G mode will drain the battery more quickly. (This is why some phones offer the option to use 2G only. Note also that some providers only support 3G on their network, so have commercial agreements to use 2G on other providers' networks. Of course, there are procedures for the phone to navigate before it can access a different network. You've guessed it – another burden on the battery.)

A rural phone, then, has to use a stronger transmit signal and perform more operations than a phone in a city. A new phone can have just 2 hours of talk time and far less cellular data duration simply due to the phenomena described above. From a user's perspective, this is a more serious matter than the annoyance of receiving the same text a dozen times. It is the expense of continually charging the phone, undermining the service's value. Far beyond 'inconvenience', the technology, which elsewhere would offer tantalising luxury, here borders on the unusable. Temperature and handshaking issues are just two examples of the many problems associated with technology that is designed for one type of user – urban residents with ubiquitous mains for charging – but adopted by others due to the potential benefits; where first world solutions are transposed wholesale to the developing world, without sufficient insight into the realities of energy poverty.

In these scenarios, regardless of the reasons for poor product performance, it is the brand name that takes the blame. And if a mobile phone only offers very short service before needing to be charged, the user is not likely to purchase a TV, computer or internet device from the same company or to recommend a damaged brand to others.

Here again is the opportunity before us: to showcase the ways and pitfalls of energy-poor markets to the largest companies; to point out that carefully designed hardware and the utilisation of local resources need equal attention; and to demonstrate the brand value that comes with guaranteeing reliability.

Endnotes

1. I use the example of lead-acid battery technology because it is the most common type for larger stand-alone PV solutions; and more straightforward than other more modern technologies. The characteristics of other battery technologies, such as lithium, can be exploited in other, more complicated ways, to obtain the same end benefit.
2. For the purposes of this text I will use 'cell' as a generic term to refer to the tower and the sectors it covers.

CHAPTER 23

Commercial growth from listening to the local market

Abstract

If our pursuit of reliability epitomises our value statement (as discussed in Chapter 22), it is market-driven product development (MDPD) *which is at the heart of growth.*

The people who are most capable of articulating commercial profit potential are those who are immersed in the market. Here are people who use autonomous energy services by necessity; who understand local customs, local perceptions of value and worth, reality versus hearsay; belief systems, cultural practices, local history, environment and language. All of these things and much more.

Local insight will unlock the commercial market and expand the profit potential that can be gained from the supply of autonomous energy services in the developing world; a potential which, currently, is held back by ethnocentric attitudes and inadequate products.

There are at least half a dozen very large opportunities apparent in the developing world at this moment. They maybe won't be as big as the PV-phone charger sector in terms of numbers, but could conceivably surpass it in terms of revenue.

The guarantee of services is essential for many reasons. It is a feature of supply that will help break into many markets. It is a strong brand asset, inspiring customer confidence and serving as a valuable element for future business.

But it is by no means the whole story. We have explored the reasons for engaging with local workforces, and the fastidious attention to product performance from the outset, in order to attain reliability. This costs money, and in itself does not justify the initial commercial expenditure.

Something else has to give. We must come back to the overriding motivator for multinational suppliers: namely, the colossal market potential in this emerging sector. If 'guaranteed services' represents our brand value, what must accompany it in the interests of meeting this huge potential, this growth? The answer lies in considering the rational need for a given product; to listen to the market. If our pursuit of reliability epitomises our value statement, it is *market-driven product development* which is at the heart of growth.

Market-driven product development

Market-driven product development (MDPD) means that people within the target customer group assess the need, demand or desire for a given service. Products are then produced to meet those requirements.

Many people assume this is the way products are always developed. This was certainly the case in the past, when a *need* for a product was primary in people's minds instead of a desire. But in our consumerist world this is no longer the case, the line between utility and aspiration becoming ever more blurred. I have laboured the point previously about products made against a price-point; this is so widespread that people seem to forget how wasteful, illogical and damaging it is; millions of items produced each year that no one needs, are unfit for purpose and rapidly destined for landfill.

Native markets, and products designed for others

In the industrialised world, market-driven product development happens, to an extent, by default. Companies, designers and developers live in the market and environments in which the results of their efforts will be used – for example, people who develop televisions tend to be TV consumers. Yes, technology drives progress. But, fundamentally speaking, design is informed by an understanding of what modern users need and wish for. For example, the multi-channel capabilities of digital set-top boxes were subsequently integrated into the next generation of televisions. When on-screen menus first appeared, they were quickly adopted across the TV manufacturing industry. The same is true of cars. People who develop cars drive them all the time, their testing grounds reflecting the familiar environmental dynamics of their target market. In an extreme case, software engineers can only do their jobs by being, in turn, users of software; which introduces further challenges into the mix when assessing the needs of customers.

Conversely, the developing world is full of equipment which is not specified by the people who live there. The technology is created by people living in industrialised societies, and it shows. The products you find in energy-poor regions generally fit into one of two categories: they were designed primarily for industrialised societies; or specified to the lowest cost, for 'poor' people. (They can of course be both of these.)

Such products are typically not appropriate for the developing world. So much of our modern equipment requires a continuous internet connection and a cool, low humidity environment. They can be complicated to use; and, above all, nearly all require stable, good quality mains electricity. But it is rare to find products which are conceived or defined locally. Consequently, a product designed for the industrialised consumer will not offer anywhere near the same value for a user in an energy-poor nation – its 'fitness for purpose' is in no way guaranteed out of context. (Remember the example in the last chapter, of the effects of temperature on battery performance, and the issue of mobile phone batteries draining more rapidly in rural areas.) Multinational suppliers have largely failed to establish brand loyalty in the developing world, and this is in no small way due to the variable performance of their products – products which have otherwise proved extremely useful and popular in their home markets. But out here, with more extreme temperatures and operational demands, 'low-cost' phones and other devices have capabilities compromised for the sale price, and largely underperform against their own limited potential in real-world practical use.

Meanwhile, there are more obvious matters to consider. A typical modern slimline television (designed for the European or North American markets) may well struggle in a dusty, humid location where it is carried by hand and knocked about. Many homes and buildings will fail to support a wall bracket (simply, the wall would fall over with the weight of a television). Instead, it is propped up against the wall and stood on drinks crates; now vulnerable because of its slim footprint and awkward centre of gravity. Elsewhere, various attempts to produce equipment outside of the developing world have, however well-intentioned, shown a truly inadequate practical understanding of what people need. The One Laptop Per Child (OLPC) initiative – which for most of the 2000s was publicly committed to providing computers to disadvantaged children in the world – was the project that surprised me most in terms of how much interest and financial support it generated. Disappointingly, it saw recipients struggling with the proprietary operating system rather than being productive; its short-lived low-cost appeal offset by its limitations and lack of compatibility with other devices or systems. The 2005 United Nations' World Summit on the Information Society saw some African representatives voice their criticism of the OLPC project; questioning the cultural mismatch of applying technological solutions to the developing world (Warschauer and Ames, 2010).

Local insight and growth

What is the solution? How can we ensure that energy services are suitable for energy-poor regions? To begin with, let us avoid the false economy of transposing a ready-made product from a foreign context, and revisit first principles. What it is we are trying to achieve?

We have said that 'Selling Daylight' in these parts of the world is only viable if we introduce something with an inherent capacity for lasting growth – a product with a clear commercial justification, in the *target* market, in the first instance. To do this, it must be conceived and designed specifically with this target region in mind, not least in matters of practical usefulness and durability. This can be achieved only by understanding the quantifiable (and realistic) potential in its use.

(This potential, if you remember, does not fly fully formed out of the box. The principles of 'Personalised Energy' stand or fall on the ability to draw value from the person operating the equipment, and the cognitive ease with which they incorporate the technology into their lives. Our holistic solution must not fail to acknowledge this.)

So this requires us to engage with the people who most understand these regions; the environments, the customers with all their needs and desires, and consequently the scenarios in which the services will be used. This process is what market-driven product development is about, and it is the best way – I would argue the *only* way – to ensure long-term growth. Each product and every service solution must be cost-optimised with the appropriate level of versatility, in order that the diverse user base will get the most value from it.

Market-driven product development sounds so obvious and mutually beneficial to both markets and suppliers; it is incredible that it is not widely practised. Within the energy sector this, quite frankly, amazes me solely on the grounds of the enormous commercial potential that is being missed. But of course, our world is not logical and our societies are not how they should be. The market focus on purchase cost plays a significant part, but so too does the way in which the industrialised world views itself in relation with other cultures. The complex reason lies both in the myth of a technologically advanced and somehow superior 'present', as well as various legacies of colonialism and the religious rectitudes of the past. Put more simply, our façade of living in an advanced society gives us a condescending assumption that 'we' know what is needed for 'them' – and how to provide it.

Having people on the ground in the developing world, on the other hand, feeding ideas to manufacturers directly, provides the essential communication to invoke deeper, more formal research and justification, with the end goal of producing a technical design specification and business case. This is such a critical issue to understand. The present market size is not limited by money or technology. Local insight will unlock the commercial market and expand the profit potential that can be gained from the supply of autonomous energy services in the developing world; a potential which, currently, is held back by ethnocentric attitudes and inadequate products.

Let me offer an example of how easily suppliers can misunderstand the value laid before them. In the early days of the PV industry, lighting was quickly recognised as an application for which PV clearly had strong value. PV was clean and could avoid the harmful smoke given off from fuel-based (mainly kerosene) and candle lighting; 'solar' was far safer as no naked flame was present; and most importantly clean solar was cheaper over the life of ownership (which should have been at least 2 years) because it avoided the continuous purchasing of fuel.

But when, finally, someone had a mind to investigate, it was discovered that the users' actual motivations for buying solar lighting, in reality, were as follows: instant control – the ability to switch on or off without the need for matches or cigarette lighters; a steady, unflickering light, which meant easier reading; and thirdly, free light was seen as a significant source of pride in the equation – the user had their own light without paying anyone, or undergoing the effort to source fuel. It was a status symbol. (The health and safety aspects were interestingly quite a way down the list, due largely to a lack of understanding.)

It is incredible to realise that the three most diligently promoted assets of generic solar lights did not match the attitudes of the buyers. Some 30 years or more after the first supply of solar lighting to rural communities, fuel offset and ownership value are still the highest metrics of promotion. It is worth acknowledging how important fuel offset, health and safety issues are for *funding* such products and for financial subsidies (many solar lights would likely not have become reality without these arguments on the supply side); nonetheless, this is a valuable lesson in the difference between perceived need and actual demand drivers – in other words, buyer value.

Understanding local perceptions

During a visit to Ethiopia in 2006, I was with two other westerners in a vehicle parked on the side of a small road in a location some 30 miles from the nearest city. A young woman approached us out of curiosity; as she came closer we noticed she was carrying a young baby on her back. We also realised she couldn't have been more than 14 or 15 years old. She had a beautiful face with radiant eyes, and a friendly manner, but just as we were starting to engage, the person we were waiting for emerged from a house and gestured that we needed to leave. It was the end of the day so all we had to offer the young girl was a half-full 1.5 litre bottle of drinking water. She looked a little perplexed as she took the bottle but gestured that she was grateful. As we drove away we looked back to see her emptying the clean water onto the ground, before replacing the plastic bottle cap and showing the bottle to her baby with a smile.

Value is a very personal thing. Perceptions of what is good or bad, right or wrong, sensible or foolish, safe or dangerous, cool or uncool, useful or unhelpful (and of course, financially attractive versus a waste of money) differ from one region to the next; or even from one person to the next. From that day in Ethiopia, I have learned never to underestimate the importance of context in understanding local perceived value. I learned that what we think of as a solution (in this case, clean water) may well be true, but those we aim to help may not necessarily understand why. Interpreting these subtleties astutely is essential for profit and growth. Just as a desirable consumer brand needs to be 'on the pulse' of its target market, an understanding of local perceptions is integral to assessing demand.

This specific event also caused me to look differently at all products and solutions supplied into the developing world. Even as an experienced PV design engineer, with what I thought was significant knowledge of market demand, I became aware of the necessity of involving local people in the specification process. In earlier chapters I mentioned a remote-controlled solar light (the Sundial), which we discovered was used by wealthy households as a reliable backup device. We determined the overall specification of this product only by observing, sitting down with families and discussing what they actually wanted from a light – and how much they would be willing to pay for it. I was surprised at the difference between people's accounts and what was available in the market. When I investigated the commercial viability of their preferred high-value product, I actually found that it was cheaper for some end users.

What we think of as a solution may well be true, but those
we aim to help may not necessarily understand why

A 'solar light' and a 'solar home system'

Going back to 2007, solar lighting solutions in the developing world fell into two general categories: small, portable lights in the $20–$40 range sufficient for reading or craftwork but not bright enough for a room; and solar home system (SHS) lighting solutions which, as their name suggests, were not portable and specifically for domestic use. SHSs were commercially sold from the 1970s onwards and for many years were the most common 'PV system' in poor, rural locations. The cost to the user was anything from $150 to over $600, depending on location and what else was supplied – TV, radio and so on. (The economics of early adoption applied: earlier installations were expensive; subsequent ones became cheaper.)

Historically, the SHS comprised separate components procured by a local company or agent, who would assemble and install them. A typical system would comprise a PV panel, a battery, a charge controller, cables, connectors, labels, light switches, the light bulbs and their fittings. Assuming all installation work could be carried out at the home of the customer (meaning there was no pre-assembly of components), this system would take around an hour to install and require a skilled, trained electrical technician, primarily due to all the wiring connections. The technician would need a vehicle (a scooter would just about suffice) and tools for mounting the light fittings and switches, cutting and running the cables, fixing the PV panel and making electrical connections to the battery. They would then have to test the system to ensure it was safe and functioned correctly.

Against this standard format, now consider a complete product solution, designed by a manufacturer who responded to locally derived market insight. The 'new' product is a lighting kit in a box. It includes two lighting units, each with internal battery and LED; each independently operated either by a switch on their case or by a remote-control unit (such as in any domestic TV or stereo). These lights have mountings for hanging from a wall or ceiling, but are small and convenient enough to be carried as torches. The separate PV panel can be mounted permanently on the roof, or taken outside each morning. Its cable is long enough to reach inside the house and its plug will visibly only fit one socket on either light unit. If the second light needs charging it is plugged into the first. The remote can also be plugged into either light, allowing all three items to be charged simultaneously. An indicator on each light and remote control shows that charging is in progress, or complete. Because the lights use the latest battery technology they can also be used to charge a mobile phone or a USB device. A remote sensor on the lights enables motion detection, either to switch on when someone enters the room, or more importantly to switch off and conserve energy if no one is using the light. Finally, the remote also contains its own low-power LEDs, so can be used as a pocket-sized torch in its own right.

Consider the brief functional description above in terms of its impact, both for the local partner and for the customer. The handy design and remote control removes the need for skilled electrical installation (so there is no wiring, fitting of switches or mounting of light bulb sockets). All installation can be conducted by the user because the items are inherently safe and cannot be incorrectly assembled. Instructions, while unnecessary for some users, are provided in the non-literary form of simple graphics. The customer has portable lights which can be used for a range of applications from reading, to craftwork, room lighting or as a torch. The lights can be used at different brightness settings to give flexibility of brightness versus hours of light. (Typically, it was found that one light would remain in the same place, such as the living room, while the other would be used for the bedroom, walking outside to the toilet, or as a general personal light.)

Because the products were industrially produced (with no dependency on the quality of local installation), they could be supplied with a 2-year warranty and a design life of 5 years. A specification based on local requirements, combined with professional manufacture, resulted in light units which can be dropped from head height onto the ground without operational damage.

This example just described is a real product, and it was cheaper to the end customer than existing solar home lighting systems of the time. On the surface, it might seem that this sort of product reduces the need for a local partner, by obviating the need for skilled installation. But the opposite is true. Firstly, the specific requirements of this product came, not from the boardroom of a corporation thousands of miles away, but from a partner who had first-hand experience with traditional lighting systems and knew the indirect costs involved with supplying them (skilled resource, logistics, user training). Also, the robust design requirements were a direct result of understanding the local context of its intended use, to include the rigours of roof-mounting, wall-mounting, repeated carrying and plugging in.

While it was necessary to identify the sales potential in order to justify the product development costs (met by the manufacturer), the generic scale of the SHS market and the growing 'solar lantern' market justified the expenditure. Meanwhile, this particular product looks very different from most; and perceived value of new products will always be unpredictable without someone on the ground promoting it in the most effective manner. The local partner is therefore also essential for sales. You may also remember that this product did not sell when sat on a shelf in a retail environment next to a selection of other, cheaper, more familiar 'solar lights'. But once our local resource proactively set about promoting its real value, sales staff continuously reported that the purchase cost was not the issue.[1]

A lighting solution is a relatively easy, but high value, product to develop. But other applications may require greater levels of insight. Refrigeration systems, entertainment solutions or consumer battery chargers have less obvious value propositions to incorporate into a design. They require more capital expenditure and have a more select customer base than the broad demand for 'light'. However, adopting a similar partnership approach reaps the same benefits, if each case successfully interprets customers' lifestyles and consumer requirements.

There is more. If you approach a given market or application as a partnership between a capable manufacturer and an insightful civilian user base, you go further than increasing the value of the service; you also stand a chance of reducing the overall services cost. Ironically, this counters the prevailing trend of compromised, low-cost-at-all-costs, performance-hampered hardware.

An assessment of spending

Earlier, in Part Five, we noted how difficult it was, without meaningful data, to analyse and estimate the market's potential value in a vast and diverse region; in particular, how much spending is 'redirected' from, say, kerosene; and to what purpose. However, from a local perspective – a community, province or even cultural region – it is still possible to assess spending habits. And this can lead to global change.

A local company's insight into existing spending can highlight obvious value propositions for autonomous energy solutions, yet they may be too small to apply the necessary time, money and resource to make a sufficiently strong business case on their own. However, were they to collaborate with a larger organisation, both parties would benefit from any subsequent expansive market assessment.

To illustrate this, we turn again to the astounding popularity of mobile phones in energy-poor regions. In Rwanda in early 2005 I was told by a local non-profit organisation that people they considered at risk of poverty were spending money charging mobile phones. The following year, while developing the business plan for flexible PV technology for my new employer, it was becoming clear that expenditure on mobile phone charging was exceeding that of spending on electric lighting of any kind. It was difficult to find any credible reports to support this, either within a given demographic or even a country. But by talking to local suppliers of energy services in Rwanda, Malawi, Ethiopia, India, South America and particularly Kenya, it became clear that mobile phone charging capability was now the number one requirement for energy-poor people. Indeed, by 2010 several reports estimated that 'off-grid' people were spending anywhere from $10 to $100 or more per year to charge their phones. The number of people quoted for this group varied, but it was at least 600 million, which in very rough terms gave an expenditure of at least $6 billion per year.[2]

Neither the handset manufacturer nor the network operator – the two organisations providing the communications – were seeing any of this revenue. More importantly for the network operator, very few talk minutes were being made; users were keeping phones switched off due to the cost and inconvenience of charging them – powering them on only momentarily to send a text. With nearly everyone on pay-as-you-go schemes in developing countries, no speech meant no revenue for the network.

In the event, it was the network operators who made the business case (long before the handset manufacturers did) for providing a charger with a phone, simply because they are far closer to the market and can see just how much a lack of energy is costing them, directly and indirectly. They knew that people had money, paying 50 cents a time to someone with a vehicle or access to a building with a charging facility for the privilege. It therefore made commercial sense for the network operator to provide a charging device. A few handset manufacturers have supplied PV (sometimes on the phone device itself), but it is the network operators who now offer separate PV charger products in serious volumes. These are either provided for free, or more commonly as a subsidised option.

The simple commercial reason for making a PV charger financially accessible is that the value of the 'solar charger' is known to exceed its capital cost. This is the fundamental value proposition for all stand-alone PV solutions. Identifying this value is critical. The commercial organisation has a business reason to cover the capital cost of the PV so they can sell their core products. This allows the user to spend little and often, without the prohibitive capital cost of the hardware.

This is one of the mechanisms by which PV-enabled energy services can propagate in poor societies. It also demonstrates the inevitability of large multinational energy equipment manufacturers entering the energy-poor markets of the world.

Care is needed here. As I have already said, it is not a given that the commercial cavalry will automatically improve the situation of energy access to anywhere near its potential. For one thing, there is a huge difference between simply throwing PV into the mix to enable a modest product range, and attaining the top table as the leading supplier of versatile, autonomous energy services. One is about selling more products. The other is about establishing an entire sector. As with all good ideas in the developing world, the risks come from the familiar cost-quality compromise, and with support.

The simple commercial reason for making a PV charger
financially accessible is that the value of the
'solar charger' is known to exceed its capital cost

The phone network operators sell their communications services through a chain of outlets, which vary in size and capability. In a city, you could purchase a phone, the expensive accessories, and receive training or advice from network operator staff; but in a rural area you are more likely to be limited to top-up cards. The customer pays the money, scratches off the silver from the top-up voucher and enters the number into their phone. A simple enough task, easily explained even by the most basic of outlets.

Reinforcing the skills of charging a phone from PV is a far more complex requirement than phone top-up, however. Network operators do not want the time and expense to handle this. The PV supplier, meanwhile, may not always have the necessary margins (having competed for the contract on cost) nor the existing networks to facilitate training on this scale.

This support model is becoming widely adopted across the world, sadly funded by the very poorest people. What should be a win-win for the suppliers and users, and an excellent example of PV enabling a fundamental service – in this case communication – remains prey to the familiar pitfalls. This demands our vigilance; a venture whose commercial future is dependent on local support in vast numbers; on the global scale, engaging millions of everyday people with a view to reaching a market of billions.

The people who are most capable of articulating commercial profit potential are those who are immersed in the market. Here are people who use autonomous energy services by necessity; who understand local customs, local perceptions of value and worth, reality versus hearsay; belief systems, cultural practices, local history, environment and language. All of these things and much more. The job of commercial organisations is to raise the bar of appreciation, that a solution exists to any energy service need, and to articulate it in commercial terms. The earlier the commercial opportunity is identified, the greater the advantage to the supplier that meets it. This doesn't mean spending millions in product development on the back of a hunch; but being prepared to gain the critical 'early mover' advantage when the opportunity mushrooms.

The best people to articulate commercial profit
potential are those immersed in the market

What is staggering about the mobile phone charger example is the sheer time it took between local organisations recognising a potential commercial opportunity (2005) and sufficiently large organisations actually taking action (2011–2012). There are at least half a dozen very large opportunities apparent in the developing world at this moment. They maybe won't be as big as the PV-phone charger sector in terms of numbers, but could conceivably surpass it in terms of revenue. I will outline one or two of these later. To find out about the others, engage with a credible local energy services company in an energy-poor region!

Endnotes

1. In the event, it was this experience which first highlighted the time and resource needed to convey value to thousands of distributed customers. The principles of 'visual information' (see Chapter 7, 'Universally accessible visual media') were incorporated into this strategy as a direct result.
2. The number of phone users has dramatically increased since 2010 (Africa alone now has 650 million) as has phone energy consumption and data usage.

CHAPTER 24
The essential role of local partner networks

Abstract

Commercial momentum is dependent on getting the basics of commercial activity right: sales, logistics and support. Within the developing world energy services sector, each one of these areas is woefully underestimated in terms of complexity; subsequently, they are severely under-resourced.

A local support network, properly motivated, provides such a resource; and is essential if we are going to deploy the quantities of energy service solutions that are commercially possible.

The 'developing world' is not a single place; each region has different issues, with tribal and national politics, religion, historical divisions, geography and climate all posing their own complexities and distractions along the path to success.

Local solar companies are therefore the navigators, the essential first line in prevention of expensive and brand-damaging mistakes.

Aside from all the areas of value that local partners hold, they can also demonstrate the value of quality services and facilitate ever-increasing sales.

It is an irony that energy-poor nations often have huge exploitable energy reserves beneath them. On the other hand, the riches of the developing world are ultimately not in the natural resources under the ground. Instead they are represented by the people, and their collective initiative. If you can motivate the people, then you have your seed for success. So far in this section, we have looked at two approaches by which we can understand the growth potential resulting from partnerships with locals. Both provide, in their own way, a commercial rationale for entering the daunting market landscape of the developing world, enabling a strong starting proposition based strictly on profit.

Increasing profit through partnership

What have the last two chapters shown us? We talk of 'guaranteed services' which underline the value of reliability, in order to deliver a dependable service. Achieving this is not cheap; but by extending the partnership with a local organisation or person, the hardware can prove cheaper without compromising reliability. A complementary mechanism exists within the market-driven approach to product development. Again, the actual hardware might be more

http://dx.doi.org/10.3362/9781788530705.024

expensive, but the cost to the local partner is reduced. In both cases, the customer – be they individuals, companies, industrial organisations or governments – purchases a more valuable service; sometimes, at lower cost. If you make this value sufficiently clear to consumers (and if you pay careful consideration to their prevailing perceptions of what this service means to them), it may be possible to offer the customer a solution at the same cost they would have paid for a less valuable alternative. The increased profit can be applied to promote continued growth, both in partnerships and in manufacture.

Growing the partnership: establishing product supply networks

What are the challenges in growing these partnerships? Some common pitfalls present themselves at this stage.

Logistics is one of the biggest problems for any supplier into developing world regions. This goes for both physical transportation and, equally importantly, the movement of information. Many organisations have tried to establish dedicated networks of trained or registered representatives for products. Inevitably this approach fails. It simply doesn't make sense. We have encountered the problems of scale, geography and diversity before; and here again, we just can't train networks to the scale needed. Renewed efforts in recent years by the World Bank and other international organisations towards 'strengthening distribution channels' (with programmes such as 'the last mile' aimed at the rural reach of distribution beyond the limits of urban centres) risk wasting resources and money. With more at stake than distribution or logistics, I would argue that it is unrealistic to train a sufficiently valuable network from scratch, even in a single region.

Distributors defection, and a suppressed market

A manufacturer of a new PV product (lighting or solar home system) enters a local market. Initial sales are relatively easy because their device has the latest LED, battery and control systems; its design is informed by lessons from the past and, although still produced to a market selling price point, the higher efficiency components offer a margin advantage.

Distribution networks are trained, after a fashion, and financially incentivised by commissions. Everything is fine for a year or so. Then a newer, higher value product arrives on the market, with lower cost or higher efficiency components, stealing the relative price point advantage. The incumbent supplier cannot compete; their sluggish research and development an inevitable consequence of severe margin pressures in the market. Unsurprisingly, the distribution network defects to the newer product. This cycle has repeated itself for over a decade around the globe, in this and many other sectors.

The hazard of low sales is ever-present. As we know, achieving quality is not cheap; local companies are not financially able to commit the time and effort to promoting user understanding and involvement. Manufacturers pursue low cost at the expense of a supportive local infrastructure. The result is a stifling of commercial sales and, at best, low demand. At worst, it damages PV's reputation. Low sales mean low margins, so local companies cannot afford to develop or grow. Tight budgets keep sales and support restricted to areas of highest population density. Sales missions then stick to these locations, despite selling products and services whose value only increases with distance from these electrified locations.

Existing local solar companies

I consider that the only solution is to incentivise existing networks financially – in a multitude of ways. The goal is threefold: first, to have these networks sell the high-value, market-driven products required by this strategy; second, to sell, support and guarantee services using these products; and finally, unequivocally, to profit from commercial growth.

Existing local companies are the priority partners for multinationals. These companies are established, in one form or another, in most of the populated regions of the world. They generally prefer to supply larger PV and thermal systems due to the respectable margins, but this project-by-project dependency is not controllable or predictable. They can also be found selling portable and consumer PV products (though from my outsider perspective, this tends to be more trouble than it is worth). Out of commercial necessity, these companies also often sell biomass cookstoves, batteries and various consumables such as electrical components. Some sell small wind turbines, or even micro-hydro, though their complexity makes them inappropriate for businesses without the requisite know-how.

These 'local solar companies' are central to our strategy. They are generally very knowledgeable of local business and people dynamics; they have had to learn them, often the hard way, in order to become established. They are also well aware of all the negative aspects of PV over the years: the badly implemented internationally funded programmes; local government 'rural electrification' attempts; poor quality solar lights and solar home systems; unscrupulous tactics of product suppliers; the 'anything is justified because we're saving the world' attitude of certain NGOs; the perils of local politics; and the persistent poor reputation of PV and 'off-grid' in general. These factors and many more have made commercial survival a constant challenge.

The treatment of local solar companies

I feel compelled to highlight the appalling manner in which some of these solar companies are treated. Despite their undeniable importance, both for present business support and future commercial opportunity, international suppliers often treat local companies as little more than a subservient delivery service. This is all-too-common behaviour for modern PV lighting product manufacturers. Not only do they severely restrict sales profits for the local company, they will have them sign a sole distribution agreement before applying price pressure on them at every turn.

When a large contract arises (often as a result of months or years of nurturing by the local company), even greater price pressure can be applied – in areas such as acceptance of training costs, or warranty support – with the threat of bypassing them and going straight to the end user. Consequently, the local sector is continually suppressed, trying gamely but in vain to function with its hands tied behind its back. The result is that users receive inadequate support, even from people who would otherwise endeavour to provide it to them.

Partnerships between multinationals and local solar companies are mutually beneficial. Local solar companies know the regional market; they have typically worked long and hard to establish themselves and develop a good reputation. They understand the native logistics, how to clear goods from customs, and how to avoid being stung by unnecessary duties. (PV products are often tax exempt or enjoy reduced import tax, but are often charged the full rate at customs, for various reasons.) And crucially, they will support what they install, because their reputation depends on it. These companies are valuable assets in themselves. If it helps, they are analogous to regional head offices for a multinational. These regional centres each have an existing network within which the partnership model can be extended: the local kiosk and small business network.

Kiosks, micro-networks and knowledge networks

The majority of poor people live in the semi-rural and rural regions of the world. Spending is day-to-day; little and often. A small bar of soap, a cup of flour, a handful of vegetables, a modest mobile top-up voucher, a single disposable battery, a single cigarette. Broadly, the outlets selling these commodities get smaller as you get further from an urban centre. Generically, we can call them 'kiosks' and they are essential for propagating energy services. They are widely dispersed and they are commercially connected to supply networks – importantly, often at their very limits. And their tentacles reach into the truly rural areas. To address poverty, energy services need to reach the people who are most remote, vulnerable and neglected. And in commercial terms, this network of micro-outlets is key to reaching the 'off-grid' regions – the largest customer base for autonomous energy services, and where those services have highest and most visible value.

If you want to bring quality services to dispersed
populations away from urban centres, the key is to
enable small retail outlets to thrive and make a profit

Kiosks are not the sole source for these regions, nor do they offer any unique services; but collectively they cater for huge numbers of people whom most commercial organisations simply cannot or will not justify serving. If you want to bring quality services to dispersed populations away from urban centres, the key is to enable kiosks and small retail outlets to thrive and make a profit.

Hundreds of thousands of kiosks and small business outlets function as sales channels. They purchase very modest quantities of goods, then sell them in smaller amounts for an overall profit. But it is crucial to recognise that they form an integral part of global sales and distribution networks. They sell the same soap, shampoo, razors, cigarettes, soft drinks and beers that one can find in Europe, Asia or the USA. The product sizes are the smallest available; but even these are recognisable to wealthier folk from our airline vanity packs and hotel bathrooms.

As you read this sentence, these same 'micro-networks' are already demonstrating significant regional, national, continental and global commercial capabilities. But, as with the multinational suppliers, we require more than mere sales; we also need a wholesale commitment to the goals of quality, value and growth – from everyone in the supply chain. This is by no means straightforward, but it is essential to incentivise these networks to become valuable assets, if we are to reach the required volume of potential customers in the developing world.

If you think about it, though, kiosks and small retailers are ideally positioned in this regard. Micro-networks can be knowledge networks, through which we can start the long and patient task of communicating and spreading our message – how we are moving away from low quality 'solar'. This alone is a priceless asset to manufacturers. They are established distribution, sales and support networks of the rudimentary kind. They can endorse quality and value when aided by strong informative visual media. They can also provide services demonstrating the commercial value of reliable autonomous energy.

These micro-networks need to see a clear opportunity to make money. They can do this not only by selling energy service products and solutions, but by themselves utilising energy services for commercial means. Let us look at some simple services that kiosks can provide – enabled by our high quality, versatile core functionality. Consider the value of renting products to those with very low spending capacity.

A rental model

The ability to rent out a PV light, phone charger or other energy service is hugely attractive. A facility within the product can enable activation and deactivation of the function either directly or over a period of time.

A customer can visit a store or kiosk, pay a small amount of money and take a fully charged PV light home with them, enabling them to charge their phone and light up their house. Its interface indicates how much energy is available, and its integrated instructions provide guidance for use. After, say, 12 hours, the unit switches off. Only the PV panel at the kiosk or shop is able to charge the product – a feature of the device's core controller. If the customer returns it by, say, 10 a.m., they can expect their full deposit back, and the owner has time to charge the device during daylight hours. The owner also has enhanced visual media instructions, which convey charging and usage best practice, and tips for informing customers (for example, that the device contains no valuable parts, and is useless after its designated operational rental period). The person renting obviously doesn't have to spend the capital cost purchasing the light; but gets more from their small expenditure than they would simply buying kerosene, candles or phone charging.

It goes without saying that the device needs to be rugged, easy to use, reliable and should last long enough for the rental income to cover its cost to the retailer. But all of its features (including the core microprocessor capabilities to allow this commercial model) are consistent with the product designs of this strategy.

Rental and pay-as-you-go schemes are not ideal; a customer loses the service the moment they have no readily available money. They are, however, the critical first step on the ladder of opportunity. The kiosk rental example also shows how commercial profit can be made by a very small retailer and the manufacturer of a rudimentary energy service solution. It also opens up options for the kiosk and small business suppliers to develop their services as revenue increases; they can purchase more devices and PV capacity, and they can diversify. Currently there are very few products on the market which can enable this rental model, even though technically it is entirely achievable. Here is an opportunity to reach millions of customers.

One particularly appealing aspect of the kiosk-based rental capability is for the smart application of aid and donor spending. Instead of well-meaning organisations shipping products from outside of the local economy without any meaningful or lasting support, they can spend the money through the local micro-business network. The physical light is given to the recipient along with a dongle or key so the light is for their exclusive use. The charging system is provided to the kiosk, along with lights to be rented out. The lights can only be charged by the partner kiosk (again, set by its microprocessor).

This approach massively reduces the need for education of the recipient users, and makes better use of donor money in terms of its real impact. Instead of distributing PV panels to, say, 20 people, a large panel is supplied to the kiosk, and permanently fixed in the optimum orientation. The single large panel will likely be around 25% of the cost of the 20 smaller ones. But the real gain is derived from using the system. As before, the kiosk owner can be shown how to keep the panel clean to ensure effective charging and perform basic maintenance. This can be taught in an hour using the system's visual tutorial.

> The kiosk rental example shows how commercial profit can
> be made by a very small retailer and the manufacturer
> of a rudimentary energy service solution

Each light will have a unique identification number, and the microprocessor can provide analytics for usage data. This can help ensure that the donor money is being spent appropriately while allowing the micro-business to provide support. This is an example of how donor money can be optimised, and data from all the lights can easily be amalgamated to illustrate transparency of spending.

This is just one of many ways in which we can deal with how the world actually *is*. Providing light to those who have absolutely nothing is a significant step in the right direction. Note that we do this by adding value, rather than stinting on cost. So, in this example, a mobile phone could be supplied with the light (and charged from it) to add value. For small-scale farmers, a simple text message telling them the present value of their crop, instead of selling 'blind' into the local market, can be the difference between being trapped in poverty, or escaping it for good.

Commercial services from kiosks

Recall what we said about having common core control systems – a generic, versatile microprocessor running the show. Now imagine that we can select the energy control peripherals against our electrical needs at a given time. Basically, this lets us increase the capacity of the system in a cost-effective manner.

A small business, starting out with a modest half-dozen rental lights can expand to provide any or all of the following: internet access and international (satellite) phone services; charging services for phones and consumer batteries (commonly used for radios); and a refrigerator for selling cold drinks, perishable foods, or even luxury goods in small quantities, such as chocolate. An increasingly popular option would be the installation of a TV showing local and international news, soap operas and sporting events. Since we are on the subject of spreading information via these extended networks, this same TV could also be viewed as a potential advertising channel for autonomous energy services, in the universal style of the instructive animations developed for the brand.

These, then, are initial illustrations of how to engage with micro-networks. They are ways to establish local partner capability to enable the goals of this strategy. So, what is its profit potential? Before embarking on this question, it is important to take a look at the basics. So many organisations enter developing world commercial markets with grand ideas. And so often, they underestimate or neglect the everyday challenges which bring such endeavours to their knees. Let us examine some of these realities.

The realities of local logistics

Logistics is an integral part of commercial activity, especially to growing a geographical presence. Delivering a shipping container of products – to, say, Kenya – is easy. Delivering the content to a remote warehouse can be complicated and costly, especially for PV. Reaching the dispersed customer base of small vendors and individuals is a further challenge. This, again, is where local partners come in.

So many organisations enter developing world commercial markets with grand ideas, often underestimating the everyday challenges which bring such endeavours to their knees

---

Distribution in the developing world has a definite structure to it. The primary regions are urban, semi-urban (peri-urban and suburban), rural and remote. These are not formally structured and vary from place to place, but they are a reasonable and useful way to picture how a product can be propagated. Existing distribution networks are the only way to overcome the challenges of geography and reach customers.

Micro-networks can deliver products to a hundred semi-rural outlets, who in their turn could each supply a hundred-fold – even if those outlets are individuals who buy a handful of products at a time. The sales at the far ends of the network are the smallest, but the cumulative total of all the rural sales adds up to a commercially significant sum. This is how to reach rural communities; the tentacles of commercial supply support and maintain these societies.

Earlier on, I stressed that, from the manufacturer's perspective, energy services have to be made to fit the format of a consumer product. This is particularly true when you look at the local supply side. Our solutions have to be truly dispersed over vast distances. As consumer products, they can be made resilient and 'logistically friendly' for delivery via existing distribution channels. Those channels do not have the luxuries of lifting equipment or even conformity to handling procedures.

Packaging can reduce the adverse effects of transportation, but it cannot be relied upon. Just a small amount of consideration can prevent expensive breakages, and protects sensitive components. It is entirely possible to deliver PV, batteries and all other components to very rural locations, but consideration of the logistics must be taken from the outset. Local ground transportation from a port or airport requires driving over harsh roads, often in ancient vehicles which were never intended for goods handling. Compared with typical use in the industrialised world, equipment and components will likely suffer excessive knocks, vibration and unusually harsh physical treatment. If local people do not understand, for example, that a PV module has a sheet of glass in it, breakages will almost certainly occur. If a battery is dropped from even a metre high, it will likely sustain internal damage that will significantly affect performance and operational life.

If this sounds obvious, consider that profits don't just come from the selling price. They come from reduced breakages and maintaining the integrity of the products throughout the logistical process; in a pioneering margin-critical market sector, attention to this can make all the difference.

For PV and batteries, damage is often not visible to the untrained eye and the product may function, but only for a limited period. (Even a PV module with shattered front glass can still function initially, but will degrade rapidly through moisture ingress and other factors.) This hidden damage is particularly troublesome as the installed PV system can still be commissioned – and paid for. It is then only a matter of weeks before the system fails.

People support

Delivering the physical hardware is just the start of providing autonomous energy services. No matter how reliable a product is, no matter how clear supporting information is, there will always be a need for humans in a supporting role, not least to address unforeseen circumstances. This, in spite of rigorous testing regimes, is the case for the most advanced technical products in the world, true for smartphones to the Hubble Space Telescope.

Consider, then, the value of old-fashioned people support, in places lacking reliable internet services, or charging capabilities. In the comfort of the industrialised world, we just find a way to get by in this regard. We ask a friend who knows, we read the user manual that was supplied in our native language, we go to the website or watch the online user video. In the field, these options disappear. Even accessing an online tutorial video requires basic literacy and English, but also reliable internet access. And the user video for your new phone might not be so helpful on a 4-inch screen in bright sunlight, as it pauses to buffer every 20 seconds, for which most people in the developing world are having to pay to download or stream.

> Local partners are the only realistic means we have to supply
> and support autonomous energy services in the local context

Even setting aside the specific need to convey PV knowledge, in the real world there is always a need for personal interaction to help overcome a problem or concern. It is this genuine support from people that encourages correct operation, reinforces value and ensures repeat sales.

Challenges and opportunities

It is a commercial imperative that our strategy proposes a way of reaching off-grid regions; the money already being poured into logistics and support is too vast to ignore. This fact alone justifies working with local partners, as they are the only realistic means we have to supply and support autonomous energy services in the local context. This is not to ignore the challenges before us. I'll say it again that the 'developing world' is not a single place; each region has different issues, with tribal and national politics, religion, historical divisions, geography and climate all posing their own complexities and distractions along the path to success. Local solar companies are therefore the navigators, the essential first line in prevention of expensive and brand-damaging mistakes.

But the difficulties in motivating vast partner networks can also turn up specific opportunities. Let us look at two such examples.

The gender divide

One opportunity lies in acknowledging that women and men are motivated differently. Women are now widely accepted as being key to any proactive change in poor regions. This in itself represents great progress, and a significant number of organisations are now specifically working with or for women. But prevailing cultural assumptions can also limit progress. Recognising the problems of patriarchies is one thing, but an increasing trend is to retreat from, rather than to engage with such dilemmas; effectively sidestepping the problem of men as a lost cause in development terms. This is a mistake.

Women make enormous sacrifices, sometimes for their entire lives, for the sake of their children, their elderly and for a cause they believe in. Men do the same, but among some cultures in the developing world this is far less evident. Engaging men is actually fairly easy – you just have to make it financially profitable. But as you might expect in the area of gender politics, there are other mechanisms at work, such as status, security and respect, not to mention the social aspects of motivation. There is a widespread but oversimplified belief that men in the developing world are fundamentally lazy. Of course the issue is more complex; I find them far more concerned with public perception and appearance, a characteristic which for us unlocks a key to the hearts and minds within a vast population. A local solar company that I know is able to motivate men to gather rural market intelligence simply by providing them with a basic solar phone charger. There are many ways to motivate and direct local people, but as with the conditions within which they live, those motivations are locally specific; and these require local insight.

Involving local fuel suppliers

In another example, fossil fuel suppliers (who, with reference to the previous paragraph, are typically men) are likely to feel under threat by anybody swooping in and providing autonomous energy services. This is particularly true of paraffin and kerosene sellers. It is the global oil problem in miniature: the spread of renewable energy undermines the core business of the prevailing controlling organisations. The fact is that clean energy services will face significant opposition in the first instance.

But again, this hazard conceals a useful resource. Local energy providers have direct access to the people in the community. Fuel distribution channels are very well established and cope with commodities far more challenging to transport, store and support than PV-enabled services. The local fuel supplier will know the areas and the people very well. They will also know how to spread information – be it valuable support, or rumour and misinformation.

Fossil fuel suppliers are likely to feel under threat by anybody
swooping in and providing autonomous energy services

Distributors, shops and individuals in the fuel sector all supply and install spare parts, and they service the equipment, however basic. Essentially, this is a ready-made technical support network, just like the one we need for our clean energy services. It seems counterproductive to become direct competitors. Instead, why not approach local energy providers, offering them the opportunity to expand their businesses? This positive engagement basically means ensuring that they make a net profit against any reduction in fuel sales resulting from PV-enabled services. At the same time, it co-opts the very forces which would otherwise hinder our progress.

A friend of mine terms this approach 'Go local, don't take away from local'. We are, after all, aiming to involve people in the supply of PV. Lighting is the primary competition to their fuel business in the beginning, so at least an initial emphasis on PV-enabled mobile phone charging, TV, internet and refrigeration should be attractive for enhancing both their business and their personal status. Like it or not, teaming up with local fuel providers takes a proactive move against potential competition and reputational aggression in the future.

The growing value of micro-networks

A micro-network, once established, is immediately valuable. It may be basic to start with, but for every month they are involved in PV and high value energy services, the more valuable they become – and the more money they can make. Commercial momentum is dependent on getting the basics of commercial activity right: sales, logistics and support. Within the developing world energy services sector, each one of these areas is woefully underestimated in terms of complexity; subsequently, they are severely under-resourced.

A local support network, properly motivated, provides such a resource; and is essential if we are going to deploy the quantities of energy service solutions that are commercially possible. Aside from all the areas of value that local partners hold, they can also demonstrate the value of quality services and facilitate ever-increasing sales.

CHAPTER 25

Global organisations partnering local networks

Abstract

It is the combination of the in-country energy services network and the international supply base – a global organisation functioning locally – which can fulfil our brief; if we consider that an international supplier can turn an idea for serving hundreds into products desired by millions.

But our strategy, if it is to take flight, also requires our local networks to grow in scale and competence, in parallel with commercial growth of the hardware manufacturer.

'Growth', in context, is my preferred term to alternatives such as 'business development', for this reason.

Among the elements constituting growth, I would suggest local market insight and market intelligence. It requires intuition of real-time market activity, understanding of local perceptions, wants, needs or desires.

In this chapter we examine some of the many ways in which locals hold the key to growth – towards that 'absolutely enormous commercial potential' we uncovered earlier.

Hopefully, I have been able to begin to demonstrate the latent power of local resource in the developing world. It is equally important that we recognise these local partnerships for their commercial value: their potential for increased revenue and increased margins, and for diversification (itself an essential requirement within autonomous energy service markets). The market insights they possess are essential to growth; and we shall investigate this further.

The elements of growth

The giant technology companies of the world, and the local supply and support networks, each share a mutual motivation – financial profit. Here, essentially, we are concerned with the many ways of creating and increasing this profit. But our strategy, if it is to take flight, also requires our local networks to grow in *scale and competence*, in parallel with commercial growth of the hardware manufacturer. 'Growth', in context, is my preferred term to alternatives

http://dx.doi.org/10.3362/9781788530705.025

such as 'business development', for this reason. And it is this broader, qualitative definition that I shall use, to embrace the idea of committing to local resource in the long term.

Our strategy, if it is to take flight, also requires our
local networks to grow in scale and competence

What, then, do we classify as 'growth'? What are the types of qualities which need to improve along with the profit line? Among the elements constituting growth, I would suggest local market insight and market intelligence. It requires intuition of real-time market activity, understanding of local perceptions, wants, needs or desires. Certainly, it demands the talent to identify new opportunities, new technologies and new energy service applications. And it requires innovation. Let us examine some of the many ways in which locals hold the key to growth – towards that 'absolutely enormous commercial potential' we uncovered earlier.

A continuous cycle of development

As we pursue the dream of ending the cycle of poverty, the cycle of innovation will never end. All the while, stand-alone PV experiences a wealth of technological advances: higher efficiency PV modules available in increasingly diverse formats (some of these without glass, foldable and wearable); lighter, more portable and more durable batteries with higher energy densities; LEDs with lower power demand and enhanced light output. The same is true for user devices and appliances, with lower energy requirements and more rugged build quality. All of these continue to improve the value, convenience and affordability of autonomous PV services.

And this, after all, is the whole point of providing access to modern energy services – to enable development. As individuals, households, organisations, regions and indeed nations advance, more and more energy services will be necessary. Lighting enables reading and craftwork. Computers and communications enable more advanced learning and business activity. Refrigeration enables food storage. And entertainment (be it live, broadcast, interactive) makes people happier, more naturally productive; collectively, and culturally, more vibrant and confident. This is also growth.

The most important aspect of local partners is to support the ever-increasing needs and spending ability of the people who benefit from quality energy services. The provision of autonomous energy services starts with the financial incentive of large commercial organisations. Local people enable the development of the business and its commercial growth. This process never ends; it just changes in form.

Entertainment makes people happier, more naturally productive; collectively, and culturally, more vibrant and confident. This is also growth

---

Market insight for innovation

Applying new and advancing technology to existing applications is commercially valuable, but the real gains are made when you identify step-changes in value and commercial potential. Innovation – recognising opportunity, seeing demand or a gap in the market – tends to come from those immersed in that market. It is the people experiencing the problems, waking up one morning thinking 'I wish I had x for this task', 'If only that family had y labour saving solution', 'Look at how many products you could sell if it did z!'

But it is rare for the resulting technology development to come from small, local companies; in our case from small local energy services partners. The idea itself can be simple, but the solution can be the opposite. Innovation, therefore, works best as a partnership: between an organisation prepared for the typically high cost of development, and somebody on the ground, capable of articulating an opportunity. Unless suppliers can identify the commercial market for a product innovation, they will not undertake the development. And if innovators are not financially rewarded, the truly ground-breaking innovations are not shared, and will ultimately never materialise.

Partnerships also bring significant legal benefits. Patents and intellectual property present considerable issues for developing world citizens. It may seem somewhat counterproductive that the patent process requires details of an innovation to be made public. In today's world, with law an expensive business, a patent becomes a liability if its originator is unable to defend it legally and commercially.

This notwithstanding, innovation in developing world countries is widespread. Even after nearly 30 years as a professional engineer, I am regularly impressed during visits to Africa and Asia by the ingenious solutions people find for every type of problem; and the ingenious uses for items which would otherwise be consigned to waste in industrialised societies. The ongoing challenge for innovators is finding access to capital, and the markets into which they can sell. By definition, solutions for the poor cannot make significant returns; this challenge is less about profit, and more a case of meeting development costs.

The money is not a problem when an international partnership exists. Rich countries spend billions of dollars on research and development. Despite this spending, they regularly miss huge opportunities. An example of this is non-glass thin-film PV. A handful of companies have spent over $3 billion in the last decade on 'flexible' or '3rd generation' PV with none of them making annual profit, let alone repaying capital. Indeed, many have gone out of business. These companies could have been commercially active if only for a different market focus; namely, if they had engaged with local innovators in the developing world countries, they would have seen the vast opportunities, and the growing demand.

Innovation works best as a partnership: between an organisation prepared for the high cost of development, and somebody on the ground, capable of articulating an opportunity

---

Having said all of this, have I not asserted previously – more than once – that this strategy can be implemented without technological innovation? Does this not contradict what I've declaimed above? The 'absolutely enormous potential' is indeed there now, and can be realised utilising existing technology. But that doesn't mean that we ignore the potential of what innovation can bring. What we can do now will form the brand and commercial turnover. Innovation through local insight, well-funded product development and early mover status fuels growth towards that gigantic commercial potential. In the never-ending cycle of innovation as described, nothing stands still. With such commercial opportunity and urgency, sitting and watching the wheels is, frankly, not an option; sometimes, the pioneers are the ones who have adopted early and got their hands dirty. With this in mind, the remainder of this chapter offers some examples of current commercial opportunity within the energy-poor regions of the world, taking place today.

Capital funding from commercial organisations

Multinational electrical equipment manufacturers are not the only companies requiring energy for their devices. Everywhere there are companies selling energy-dependent commodities, such as cold drinks, foodstuffs, medicines, and film and video content. The international suppliers of these commodities generally stick to urban areas, simply because that is where grid electricity is. Any significant rural or semi-rural presence they have will only be on mains electricity spurs. But once companies like these begin to grow more confident in the reliability of a guaranteed autonomous energy service, there is a very attractive asset for them to use; and such limitations begin to fall away.

A cold drink

Take the example of the major cola drinks companies. In a typical industrialised world convenience store, you are likely to find a tall, glass-fronted fridge, exclusively displaying products made or distributed by the corporation who provided it. They did this because they sell two products: the drink and their brand. Their brand position is such that they always want to sell the drink cold, incorporated within a prominent continuous advertisement. This same approach is taken all over the world by beer companies, dairy suppliers and ice cream brands.

Now consider a location without mains electricity. A few years ago, a solar company run by a friend of mine was asked by a local beer producer to investigate PV-powered fridges for remote locations in central Africa. For comparison, they were provided with a typical fridge – AC powered, with a big glass door and bright white fluorescent lamps inside. Taking the holistic approach, they tested, in parallel, an energy-efficient (DC) chest-style fridge, with the same storage capacity and the same continuous temperature for the product.

A chest design is vastly more efficient, simply because cool air naturally sinks, so it is mostly contained when the top lid is opened. Further energy-efficiency measures go to make some significant savings. After testing energy consumption for a week and regularly removing beer (purely for test purposes, you understand), the figures were used to make PV system proposals. In a shaded but external location, the standard fridge costs around US$45,000 to power reliably for 5 years, whereas the chest design costs around US$4,000! This small reimagining of the problem shows that it is not cost-effective simply to replicate the standard sales format of big, glass-fronted fridges commonly found in industrialised societies. The autonomous chest fridge format is commercially viable.

The perceived problem here is branding. The energy-efficient fridges described above are big, opaque, trunk-shaped boxes, and much bigger physically than their counterparts (though, notably, not heavier thanks to light-weight insulation). However, by including an illuminated logo on the vendor premises or kiosk, you will make it visible at night from literally miles around. Modern LED systems for display boards, active logos or advertising are extremely efficient, with less than 10% effect on the energy consumption in the above example. The question is: what is the monetary value of being able to provide cold drinks and branding in a location with no electricity? Simply, *people drink more beer when it is cold*. The beer company undertook the investigation for autonomous power after establishing this among rural people, who, when first presented with a cold beer, preferred a 'cool' one (warm by any industrialised society standards), but were converted thereafter. The principle is well known and, unsurprisingly, the same for soda and other consumer drinks.

Businesses meet the capital cost

This is a win-win situation for consumer suppliers and those with low incomes. The consumer product supplier meets the capital cost of the enabling equipment, so that they can sell more of their product. That product is accessible and affordable to low-income populations living away from the grid – over 1.3 billion people. These people do not have to buy their own fridge, nor are they excluded from the end commodity due to cost or location; they spend little and often, and locally – just as they always have.

For anyone experiencing unease at this point, let me interject on a personal note for a second. The example of cold drinks is deliberate. I am not a fan of them, nor do I espouse their consumer culture, their environmental impact (several times the water volume is required per finished product volume) or the health issues they represent. But these products are in demand, and in no small part due to their association with our consumer world. To acknowledge this is to concede one of the contradictions in the relationship we conduct with corporations, and our assumptions regarding remote regions of the world. What I have learnt, and what this particular contradiction means for us, is that this same efficient, fit-for-purpose, autonomous fridge can be used for storing medicines and vaccines, in regions where health provision needs are most acute. They can be used to store dairy, meat and other perishable products – a saving in time and money, and a boost in terms of hygiene and income. Be in no doubt that this will only be realised with the clout of corporate players, in partnership with those in the field.

Moving on, and in purely commercial terms, this commercial model for cold drinks is driven by the same motivations as the mobile phone network operators: pay for appropriate equipment to be put in place, so that you can sell more of the company's core product. The financial benefits in each case are entirely representative of the commercial justification for dozens of commodities and services. A PV charger (with its own battery) is the most versatile and useful phone charger for rural locations. A PV-enabled refrigerator fits the bill likewise. Finally, this commercial rationale is not limited to new, off-grid regions. Commodity suppliers can also enhance their commercial position in existing, grid-enabled areas where their brand might be compromised. The obvious example of this is where a refrigerator is plagued by unreliable grid power, where the drinks are rarely actually cold.

The example of television

A multinational autonomous energy supplier can use the commercial value of local networks to attract user device manufacturers. A pertinent example is televisions for the developing world.

A few years ago, I compiled a specification for a TV manufacturer. I was fortunate to work with local people who were sufficiently technology-aware to see that the existing products did not merit their hard-earned money.

The supply of televisions happens to be a rapidly growing market in energy-poor nations. It is driven by the availability of digital media (on a USB device, disk or downloaded) and the increasing availability of satellite and local television content. In particular, the TV market has boomed since the widespread availability of flat screen devices. Unfortunately, much of the spending to date has been on TVs designed for developed societies, most of which are unsuitable for local transportation or challenging environments. Some only work from a battery-dependent remote control. Others support limited file formats, defaulting to the region for which they were originally intended.

In cases such as this, where there is growth in generic demand, a local partner network could provide initial market feedback to their multinational partner who, in turn, would assess the market's potential for that end customer service. Out of this emerges a business justification to develop the hardware. *This was precisely what we experienced*. The commercial opportunity we uncovered proved convincing, to the point that a low-power TV became a standard option with several solar home systems.

Televisions are a perfect example of the need for the market-driven product development we saw earlier. The commercial manufacturer is uniquely positioned to seek opinion and market data from its sales regions around the world. They have the commercial prominence to speak to key stakeholders in local areas – broadcasters, advertising agencies, governments, and film and media organisations. And, ultimately, they can tailor a product specification offering the greatest versatility against the inevitable breadth of requirements.

How to attract a TV manufacturer

The autonomous energy manufacturer already knows how to make quality, high value service solutions. They can use this knowledge to select the most appropriate user device manufacturers.

The approach they take is critical to this relationship. The usual discourse – 'Please advise us on which is your lowest cost TV', or even 'which is your lowest energy TV?' – while simple and effective, nevertheless risks undermining the value of the genuine market knowledge being brought to the table. A better approach, which references the background to the product development, would go along the lines of: 'Accept our guidance in how you produce or tailor your televisions, because we have verified demand with our local partners. We will then provide you with a market of a billion people currently out of your reach.'

What is the nature of this guidance? Primarily the TV should be as energy-efficient as possible to reduce the cost and size of the PV supply, with no power conversion between the PV system battery and the TV. Valuable guidance is also informed by understanding the market: for example, support for multiple video formats and a USB port. In addition, it must be resilient to humidity and extreme temperatures, while easy to carry by hand and able to stand on an uneven surface.

'We will provide you with a market of a billion people currently out of your reach.'

The remote control, meanwhile, should be rechargeable from the PV system and the aerial (either supplied or an optional extra) should be of proven quality and durability. All hardware must have a minimum operational life of three years and minimum warranty of one year.

Most importantly, we require universal accessibility. All user interfaces must be clearly identified with universally meaningful labels. Hardware and supporting information must be visual, requiring no reading whatsoever; and produced in accordance with the visual instruction system we provide. Following all of these guidance instructions will result in a reliable media player, which functions predictably, can be used in most places in the world and will last for at least three years without operational costs.

Adapting an existing device

Ideally, a device such as a TV could be produced specifically for the developing world market, and we are starting to see this. This affords a greater degree of brand congruence, such as embedding features required for the rental plan, or handshaking – an option to make the TV exclusively compatible with the PV solution supplied.

More likely, in reality, is the practice of adapting a standard, existing device. But this too can be tailored to increase value for the customer. To start with, the internal power conversion from mains electricity to DC can be removed (the PV system, like most equipment, is already DC). This is technically very easy to do. Even simple settings can make a difference; for a TV, the optimal balance between energy consumption and screen brightness can be achieved in software, rather than with manual switches. Packaging can be adapted to provide greater protection, while simultaneously forming a stand for the TV.

There are many ways to make an existing product more appropriate for energy-poor customers if one has sufficient insight into requirements, the nature of the user and their environment. While this feedback from local partners to TV manufacturers is growing in momentum, it hasn't reached our typical grid-focused multinational energy suppliers quite yet. However, we are now seeing 18-inch flat screen TVs with as little as 10 watts of energy consumption, and 32-inch TVs running as low as 20 watts. (For perspective, a 32-inch TV from a typical high street supplier will be over 50 W for an LED type, maybe 70 W for an LCD and over 150 W for a plasma. An old-style CRT TV would have consumed over 100 W just for a 22-inch unit – the biggest that could likely have been carried by a single person.) Remember, the cost and size of a stand-alone PV system is directly proportional to the energy required by the device.

Overcoming capital cost

This business strategy, remember, aims to deliver value at the point of use, by virtue of joined-up product development, innovative design programmes, pioneering non-literary communication and commitment to quality over lowest cost. This, of course, is an expensive undertaking, especially when our products are visibly not the cheapest in the market. Overcoming the capital cost of hardware, then, is as relevant to the PV local partner and other businesses as it is to individuals. Obviously, wherever someone else can meet the capital cost of hardware, local businesses in relatively poor areas are going to benefit. But they are too small to attract commodity suppliers on their own.

The multinational energy services supplier, on the other hand, is large enough to make such strategic agreements: with manufacturers of consumable products requiring energy for storage (cold drinks, medication or food items); or for operation and recharging (TV, digital cameras and so on). Or in fact both. Casting wider, movie production companies can purchase autonomous TVs, Blu-ray players, projectors and other devices in order to sell the commodity of a film.

Here is where the giant reach of the market potential becomes manifest. Partnership capabilities, combined with brand and reputational development, enable access to ever-greater commercial budgets and larger contracts – national infrastructure projects such as government offices, healthcare networks, phone and internet networks. Viewed together, these mechanisms contribute to make reality of the 'absolutely enormous commercial potential' we have uncovered.

Growth and progress

Placing growth at the centre of the business partnership helps focus on the details, particularly decision making. It is understandable to be cautious over issues such as employing sales and support staff, choosing the correct equipment for logistics, or providing proper levels of support. Such decisions are easier to make (rather than defer) against a clear underlying direction of long-term growth, and an explicit value proposition of reliability.

I will continue to make this point: the market is currently held back, not because the market is anywhere near saturation point, but due to the drawbacks of products and their support. Development thrives on continued support. A brand's reputation, similarly, depends on a firm foothold and a continuous presence in a given region. This happens to be true for national infrastructure as well, if it is to expand. When the business is focused on growth, the nurturing of local people – of partners – is effectively written into the plan as plain good sense. This logically bypasses the familiar temptations to cut corners because it 'costs too much'.

And in the long term, the hope of growing the business requires new variations of value products, new and larger energy services, new applications, in new sectors. Most importantly, it demands access to new budgets. Better products will increase the scale of our market potential, for certain. So, too, will a value-added, mature network of knowledgeable, well-supported regional partners.

At the same time, companies and international organisations continue to spend money and effort developing local distribution. So I will reiterate here too: A partnership is a bi-directional relationship, not a distribution network. Ultimately, the objective of growth is a local energy services network in every region. Once such a network supplies reliable energy services (that is, more than just power supplies and products), it will attract all manner of other credible benefits. The cold drinks and the TVs I have described are just two examples of a vast horizon of commercial opportunities for generating high volume sales – all facilitated by the marriage of global business power and local partner know-how.

Financially retaining local partners

By now, we can be in little doubt of the power available in global-local relationships and their resulting networks. But unless we fully understand the mechanisms to meet their costs, how exactly are we to go about reinforcing or even retaining this value?

As with most other aspects of this strategy, there is no single funding source; and no single solution to the problem. Money comes from diverse sources, some of which are offset costs. For example, the hardware cost savings which result from guaranteed services and the profits we recoup from our smarter, market-driven approach to product development. Also, funds are calculable from service agreements with commodity suppliers (cold drinks, mobile phone network operators, advertising and so on), from user equipment suppliers (televisions, refrigerators, internet providers) and more.

A partnership is a bi-directional relationship, not a distribution network

Remember that all of these services apply to every potential customer group – individuals, companies, governments and industry as a whole. In many cases, even a single support agreement is sufficient to cover the costs of the local partner. This can be a drinks supplier contract to guarantee the all-important requirement to keep the drink cold. It can support the reliability of the TV network with a view to increasing advertising revenue. Alternatively, it can cover internet, mobile phone, healthcare or banking networks.

This is more than just conjecture. Service contracts for industrial off-grid installations have been around for decades and, when the equipment is designed appropriately, these contracts are decidedly profitable.

Thousands of support personnel are employed in the developing world for telecoms, air-conditioning, logistics (the far reach of DHL, FedEx, among others) and cars (breakdown and brand-specific issues). The telecoms sector employs an army of workers simply to service and continually refuel diesel generators. But for their PV-enabled replacements, servicing once per year is fine; and you get the additional luxury of pre-failure notification. Minimising this cost means keeping the supply chain as short as possible between the manufacturer and end customer. Local solar companies can supplant the need for regional importers (this is doubly satisfying in terms of cost savings and in enhancing multinational-local networks). Controlling the logistics of goods and people in a local region can itself go some way to meeting the costs of the local partner network.

Many aspects of support are without cost, and remain intangible. Simply backing local solar companies in tenders, enabling them to promote the manufacturer's capabilities within their own sales offerings, can prove valuable in its own right, promoting the partnership rather than just the local company capabilities. Nonetheless, any party involved in the supply and support network for a given region must make a profit. Depending on volunteers or aid organisations is a short-term answer, which will only cause disruption.

The goal of global local networks

Taking stock, then. All of the recent discussion has been primarily about the developing world. But the goal of this strategy is that these hundreds of local partner networks should form a truly global capability, with micro-networks in each region forming the global web for the commercial multinational. And here I mean properly *global*, not simply the industrialised world's complacent subset of easy-to-reach, dependable grid-powered, internet-ready outposts.

Paramount to all of this is motivation. We need to incentivise everyone financially, from the manufacturers of the solutions, the local solar company, the kiosk or small business, all the way through to the teenager selling a single product each day on his or her bicycle. Those teenagers, living in poor rural communities who have a friendly manner and an entrepreneurial flair, are essential to the future of the partner network. And to reiterate, the key word remains 'partner'. Profit must be shared with locals as if they are partners, not as if they are passive contractors. Local people need to be retained for the long term, and they need to be nurtured. Funding the micro-network is obviously no small task, but it is already a reality within commodities and other sectors, and there is every reason it can be so for energy services.

PART SEVEN

Why this strategy can work

A critical look at the Selling Daylight proposition

Abstract

This chapter begins with a brief summary of the whole Selling Daylight business proposition; a global strategy to personalise energy.

We then take a reality check and discuss some of the many issues not highlighted in the preceding chapters.

Does our argument have logical elegance on paper, but falter in the grimier realities of the day-to-day world? Might all of this appear a little politically naïve or disingenuous?

Why, for example, have we omitted politics and religion from this picture, given the seismic and often bewildering transformations in public life that we have witnessed in recent years?

What about alternative technologies, including renewables and hybrids?

And why are certain supporting arguments omitted? Why not leverage the environmental benefits of PV against the horrific impact of fossil fuel and the desperate state of ecosystems?

We have come a long way so far in search of Selling Daylight – a realistic vision of the near future in which our use of the sun's energy can be personalised, durable, free-standing and reliable, and capable of breaking the cycle of dependency and energy poverty in the developing world. It is a high concept indeed; it has prompted us to examine aspects of our own behaviour, our motivations and aspirations, so that we may better understand how to bring about positive change.

It has been necessary to question many of the values we take for granted, in a rapidly changing world where humanity's husbandry of natural resources is likely to be transformed in the coming decades. Specifically, we have looked at the myth of central grids; how widespread denial shores up their fragile commercial premise because, to borrow a phrase, they are seen as 'too big to fail'.

We have discovered how stand-alone PV systems are damned by faint praise as an 'alternative' to grid electricity, when in fact their benefits represent a stronger proposition than the grid itself. We have also looked at their troubled history, and identified that the key to their reputation lies in the twin watchwords of quality and value – a comprehensive rigour in conception, design, development and deployment, ensuring that a product may be

http://dx.doi.org/10.3362/9781788530705.026

fit for purpose, and therefore commercially viable. Not least of the insights we have gained is the need to communicate this value as an intrinsic part of the service we deliver, using an intuitive, visual language of instruction and support.

We have taken a deep breath and examined why poverty is still permitted to exist in a rich world, where neither money nor technology are barriers to progress. We witness with frustration how the scale and diversity of peoples, regions and cultures make the problems appear insurmountable, and thwart the most committed humanitarian efforts. But by the same turn, we have lifted the lid on human motivation, looking at the relationship between commercial reward and collective endeavour.

Finally, we explored how to turn this into a tangible business strategy; finding that the traditional models of developed nations find curious pitfalls in the emerging markets of the energy-poor world. Here, we saw the counter-intuitive notion that remote, high-risk, low-margin sectors can, if viewed as a whole and with sufficient capital investment, become the biggest market opportunities in generations.

A reality check

One day, everyone will use solar energy. This is inevitable, whether it comes about gradually, as technology, attitudes and commercial viability all catch up and sing from the same songbook; or otherwise by necessity, as traditional utilities undergo disruptive failures through shortage or misadventure. Ultimately the true worth and benefits of silent, emission-free, efficient energy generation from the sun's free resource will prevail around the world, as a serious contender for people's needs.

The truth in this statement is more complex than it appears, given what we know about the entitlement-laden energy demands of the industrialised nations, the might of grids, vested interests and the curious motivations of people and societies – both personal and political. You would not expect a shift as fundamental as this to appear from nowhere, overnight; and civilisations tend not to turn on a sixpence. But my confidence is based in everyday reality. Consider a suitcase with wheels. Once you have used one, why would you ever buy one without this innovation? (The 'how did we ever manage before' effect is familiar to anyone who marvels at the obvious simplicity of small innovations, from the patented sandwich packaging that made a billionaire, to the inventor of cat's eyes in the road.)

On a personal level, PV will become more widespread and integrated into our everyday lives when it becomes more practical. If your day bag, backpack and even clothing generates and stores energy as you go about your every-day lives, and you are able to charge your phone or other device should you need to. Assuming the solar charging facility causes no inconvenience, looks stylish and has minimal cost implication, why would you choose an item without this facility? If that seems a little distant as a concept, consider that all satellites are solar powered. Anyone who uses GPS, satellite navigation or satellite TV, is already using a PV-enabled service.

Any number of causes could catapult this strategy into a global reality. They range from underlying, slow-burning global social or environmental trends, to the chaotic behaviour of markets. Equally, a crowd-friendly viral 'like' for a breakthrough product, a sudden increase in fossil fuel prices, or an electricity failure in one of the world's major capital cities; any of these, alone or in conjunction, could change the game overnight. (It is worth noting that the price of solar energy has already become cheaper on average than electricity generated from coal.)

Nevertheless, this is still a bold assertion. Can we really make such a claim, since it rests on the actions of investors and influencers engaging commercially with multiple sectors in some of the more unstable places on Earth? Does our argument have logical elegance on paper, but falter in the grimier realities of the day-to-day world? Might all of this appear a little politically naïve or disingenuous? Indeed, a global strategy to personalise energy prompts many broader questions that I have omitted from this book, mostly for reasons of space, or in the interests of a single narrative.

Were you to analyse this business strategy formally, you might concede that I have waxed lyrical about its strengths, mentioned the weaknesses of PV's reputation, and outlined the commercial opportunities in various scenarios and environments. But what of the threats? Why, for example, have we omitted politics from this picture, given the seismic and often bewildering transformations in public life that we have witnessed in recent years? What of the rise of electoral populism, of nationalism, and the cultural and economic instability they herald? What of the putative freedoms of the press amid the narrowing of traditional political discourse?

While democracies drill beneath their own foundations, how do our development goals of sun-powered self-determination sit with theocracies and the rise of religious fundamentalism in many parts of the rich and developing worlds? What of organised religion in general, as it stands more influential than ever as a driver of peoples on a global scale? To put this in focus, how are multinational companies to be convinced that getting their hands dirty in emerging markets will yield worthwhile returns, especially in volatile times?

Elephants in the room

On the subject of corporations, does the practice of 'Selling Daylight' seem like too tall an order? Does building to a quality rather than a cost demand that we re-set our commercial boundary conditions altogether? And, heaping on the scepticism, what is this talk of 'value', linked inextricably as it is to abstracts such as 'quality', and indeed, 'virtue'? Is this just an old-fashioned moral sermon exhorting the righteous to change the world?

To begin with, we must acknowledge that private ventures on such a scale necessarily engage with, and impact upon, public life and its governance.

They do not work in a vacuum. But while they cannot be immune from politics, and they are certainly not *above* politics, they are nevertheless conceived to function on scales and with speeds that centralised bureaucracies cannot.

Certainly, 'Personalised Energy' aims to empower individuals and groups with their own means of getting by. It is predicated on the idea of private enterprise reaching homes and families otherwise disenfranchised (economically speaking) by nation states in large land masses, island states or those living amid civil unrest. And I consider this a realistic goal, irrespective of partisan tribal or political loyalties.

The realm of politics continues, meanwhile, with a largely self-seeking agenda. In the developing world, governments and bureaucracies tend to be all talk and no action with regard to energy services. What is worse, the involvement of a local politician is like a curse on that project. Undermining the whole political movement is the reality that a credible, hard-fought policy for a particular sector, region or people can often have so little meaning. Policy is an intention, a piece of paper to be referred to, but too often impotent, circumnavigated or plainly ignored.

On matters of belief, we at least addressed the notion of an inclusive and culturally neutral visual instruction system, in order to reach the greatest numbers of people. This is one of the guiding principles for anyone trying to convey the technical workings and the benefits of plugging in our products. Elsewhere, though, organised religions and other prevailing ideologies succeed in perpetuating robust power structures and biases which are pretty much impossible to crack. Realistically, no corporate enterprise can hope to.

Religion and its missionaries have a great deal to answer for. When I see a white American man with two Rwandan boys wearing ties that read 'Jesus has saved me, will he save you?' I feel a mix of anger and sadness. The social benefits can be very real, but the damage caused by religion in the developing world is undeniable. Scaring the poorest with horror stories of what will happen after they die if they don't do what they're told (and having them part with money for the privilege) places a burden that spans their lives, and those of their children. This is particularly troublesome when the neighbouring community has been told a different story. It's a lethal mix, and yes, it may threaten to destabilise your best efforts to improve people's lot in *this* lifetime.

But religion isn't going anywhere soon. Like subservient politics, it is a base condition under which we must function. With no disrespect intended, churches become customers like any other. If they bring people together in groups, they deliver punters to us by way of the PV array on the roof.

'Personalised Energy' aims to empower individuals and groups with their own means of getting by. I consider this a realistic goal, irrespective of partisan tribal or political loyalties

---

Remember that all grand aspirations and business plans have faced these same challenges since time immemorial. Earlier we acknowledged the undisputed triumph of capitalism as a model for growth, and perhaps the apparent folly of calling its wisdom into question. But 'Selling Daylight', or 'Personalised Energy' does not signal a demolition attempt on capitalism. Quite the reverse. Rather, it is a reality check on some of the prevailing assumptions of 'western' capitalism as it functions in the post-industrial world – for one, the notion of stacking high and selling cheap. We argue that this is an error when applying it in the developing world, where it constitutes not only environmental irresponsibility, but just plain bad business sense.

Level heads and ego trips

Remember that this strategy is far less complicated than some of the movements under way at this time: artificial intelligence, stem cell research and space tourism, to name just a few. Notice, also, that this strategy doesn't have a great deal to do with revolutionary thinking. Nor are its principles exclusive to solar power, or, for that matter, energy poverty. I am talking about matters of process, of integrity, of discipline – about doing something well, getting it right, wisdom in planning, seeing the bigger picture, getting the optimal outcome. This is why we have considered, at some length, the minutiae of procurement, holistic design, user interfaces and, to a lesser extent, the pitfalls of risk analysis, and the need to invoke communication theory to get our message across.

These are not ethical appeals. They are the means to re-assess the complex challenges of the modern world with shrewdness and a level head; effectively, an engineer's solution to a human problem. And while the troubleshooting, pencil-behind-the-ear approach may seem a little sterile amid the euphoria of billionaire tech wizardry, self-driving cars, virtual reality and the 'Internet of Things', they share a common ground in that they meet the challenges of the age with ingenuity and hard science.

So, for all the talk of 'quality' and the goal of tackling energy poverty head-on, this is no moral crusade. On my way to developing this book, I had reached a point where I could articulate why pretty much any solar energy product or solution would fail. Well, nothing has changed, and my scepticism regarding the potential of this strategy is no different. But my brush with cynicism led me to believe that money, rather than altruism, was the only viable motivator for millions of people to make a practical and sustained effort.

By now, I would hope that, if you saw a link to my Facebook page, asking to 'like' it, it would seem distinctly out of place. For this plan requires people engaging with real-world projects: it is not an appeal for admiration, nor a popularity contest, nor an artificially crowdsourced social revolution; at least not in the fortune cookie, happy-cult sense engendered by social media.

Because this is not primarily about saving the world. Neither is it limited to creating new technology or boasting the best energy products. It is about showing where profit can be made, with emphasis on long-term potential over the quick conscience-stroking gesture, however well-intentioned. The traditional bad guys, the dictators, the politicians, the accomplished manipulators, bureaucrats and so on, are easy targets for our blame. This diverts attention from some truly concerning behaviour. Controversially, perhaps, I believe that the greatest barrier to any strategy that seeks to address energy poverty are the incumbents – the people and organisations publicly committed to 'poverty alleviation'.

Counterproductive practices, obstructive organisations and damaging individuals abound here. So far, for example, I have been restrained in my criticism of the World Bank. But seeing their name against a given project alerts me that it will yield little that is lasting or positive. This is very sad, given its truly global presence and potential, and the sheer amount it has spent in the developing world (probably over $1 bn just on solar PV). But its appalling track record, compounded by global condemnation from those who have worked with this vast empire, testify to the enormous damage it continues to have on development.

We humans do seem to have issues with our self-importance. It is hard to overstate just how much arrogance exists in the aid/donor/development/poverty alleviation sector. And this fallibility hinders progress perhaps more than anything else. Eminent climate scientists are coerced into promoting inappropriate energy solutions to governments and to the public, riding the wave of attention bestowed on them. CEOs of solar lighting companies speak with plums in their mouths to gain funding for truly awful products, to be sold at high prices to the poorest customers. And on every trip to a poor nation, in every meeting and discussion group, are young individuals passionate to make a difference, blindly and confidently applying the wisdom of their 15 minutes of real world experience.

The rhetoric of this sector is top-heavy with superlatives that brook no doubt ('Of course we know', 'It's obvious', 'It's been shown'), but whose lack of supporting evidence flies over the heads of their uninformed peers. Self-absorption drives their continued stubborn and hare-brained adventures, and their defensive refusal to reflect when confronted with conflicting information – often a parroting of baseless assertions bordering on dogma. ('Of course, PV isn't a solution for cities.' By all means, please demonstrate why.) Meanwhile it is so telling of human nature that people spend thousands of dollars and significant effort in 'doing it themselves' rather than researching and quietly supporting proven third-party initiatives. Our delusions of grandeur, as individuals and for our ideas, make us prey to the soundbite, the good copy, the vain appearance of achievement. The successful projects are nearly always the ones that started locally and grew; whereas a headline objective to 'end kerosene by 2020', 'provide a million clean solar lights' or even 'electrify 10,000 homes' is unlikely to meet the claim.

This is why I have placed such emphasis on the broader picture, the importance of local partners, market-driven product development and the long-term mindset. It is why we start by targeting those already spending money on energy but who are denied the quality of service. It is why the strategy initially defers focus on 'off-grid' people in the interests of growing the brand, while understanding that they are still very much in our sights as the ultimate target beneficiary. It is this apparent contradiction – operating globally in order to execute the strategy, while thinking locally in order for it to take hold and function, thinking large and small at once – which I believe will keep us shrewd and mindful in our planning; and, possibly, defend us against the hubris and vanity that plague the development sector.

This all needs a touch of balance. So many organisations are involved in the provision of modern energy services, let alone the development sector as a whole. The e-newsletters and publications devoted only to solar for development would suggest there are over 1,000 organisations within the solar energy supply, sales and assessment sectors alone. In fairness, I feel the need to stress just how good some of these organisations are. Their level of insight into this complicated sector is invaluable. And efforts to keep the sector informed and working towards a common goal are to be commended. For my part, I hope for constructive feedback, for counsel in what I have omitted and what refinements should be made.

Our delusions of grandeur, as individuals and for our ideas, make us prey to the soundbite, the good copy, the vain appearance of achievement

---

Alternatives and the environment

Our solution has focused pretty much exclusively on solar PV, particularly the stand-alone variety in comparison with grid utility. Again, in the interests of balance, it merits to state that we have not discussed alternative solutions – even those which could benefit from some of the core elements of our strategy. I'm thinking here of visual information, knowledge sharing, holistic design and procurement reform.

It may come across that I think PV is the solution for everything; but it's not that simple. Early on we noted that, in an anatomy of human needs, the basics of food, water and sanitation are most essential resources for life. We have explored a business inventory of perfectly feasible solutions served by PV: supporting everything from industrial-scale irrigation, to domestic lighting, to emergency shelters. But electrical solutions, however they are provided, cannot address many of the needs or wants of the poorest people. Furthermore, and once again with a firm check on reality, there is rarely an overarching single-technology solution. Hybrid systems offer a greater range of potential, particularly in utilising local resources. These projects also go some way to promoting local ownership, because they are appreciated in the context of solving problems, rather than showcasing or endorsing a single product or technology.

All the sustainable technologies, energy saving techniques and even fossil fuels must play their part in the overall solution to the world's energy poverty and insecurity. Stand-alone PV systems are the generic platform of choice because they are the most versatile and adaptable energy solution, and as such, the easiest to describe in a global value context. Even so, the challenges are still very complex and difficult to explain. The value of the various alternative technology options may yet be defined by local natural resource, specialist skills, significant operation costs, capital finance or political support. Keep in mind, however, that most of these alternative options cannot be 'personalised'.

There is another area of discussion that might surprise some people by its absence. Throughout, I have stated the personal benefits of stand-alone solutions, or their commercial potency. Really, it is only in passing that I have asserted PV solutions as silent, clean and emission-free; but its environmental footprint has not formed a principal pillar of the argument.

It's not as if the comparison can't be made. For example, we haven't discussed the huge environmental impact of extracting fossil fuels (including uranium for nuclear power). Or, the massive infrastructure – wellhead platforms, oil and gas rigs, pipelines, the biggest ships on the oceans, refineries and storage. And we needn't stop at the extraction and logistics impacts. Electricity from fossil fuels uses enormous amounts of water, a precious commodity in the regions of interest to us. This alone is a show-stopper argument. The continuing cost of subsidies for fossil fuels – including kerosene in certain poor countries – is another.

All the sustainable technologies, energy saving techniques
and even fossil fuels must play their part in the overall
solution to the world's energy poverty and insecurity

Again, the reason for bypassing the charged and emotive ethical debate about the environment, despite the scientific confidence in man-made climate change, is that it does not figure among the primary motivators for people to get up and do something about the world around them. As I set out in the discussion about poverty, I am interested in a strictly *commercial* model for change, which is based on the likeliest spur to action – namely, profit. The environment, even in its parlous state, would be a distraction from this strategy.

But by mentioning the negative elements of the fossil fuel sector and the advantages of clean generation, it would be contradictory if we failed to address our own environmental responsibility and raise our equivalent negatives: the potential environmental cost of mining rare metals, battery manufacture and the disposal of plastics. While this is true of any high-volume supply chain, this cannot be used as an excuse. It is also dishonest to divert the blame to the high-tech smartphones or computers (rising significantly in the developing world) charged by our systems; praising their positive social impact when in the industrial world they have the longevity of fashion items.

Finally, you will have noticed by now that there are few references to *specific* consumer technologies outside of the computer and the phone. These will be transformed over the coming years, let alone decades. Artificial intelligence in our devices, truly global connectivity, medicines, translation software and image searching are just some examples of the coming changes to our lives. Because most of these society-altering technologies are dependent on electricity, there is a strong justification for associating generic stand-alone PV, and this strategy, with their emergence. But again, I am cautious. It would grab headlines and raise money, but it is a distraction with an indeterminate lead time before positive impact might happen.

For this reason, the strategy of this book is designed to utilise new technologies and resources without being dependent on them, to gain where appropriate but to do so using the foundations of proven concepts and technical solutions.

Long-termism

For all the explanations and disclaimers and 'what abouts' you have read here, stop and consider that we are attempting change, on a grand scale. A change in how people perceive need. A change in how people appreciate what energy enables in their lives. A change in what people mean by value. A reassessment of their own motivations. Commercialising change is not a trivial matter.

And how long will this change take? Doing things differently and thinking differently is suspicious, scary and 'other'. Humans are not exactly known for embracing change on a personal level. Especially where risk or potential loss of face is involved. Realistically, no amount of capital, no matter how well structured the strategy, will address such a large and diverse set of challenges in anything that can be thought of as 'short term'.

Money still drives the world, and it motivates people as individuals and groups to feats of ingenuity, with an eye to the long-term prize

---

To drum up investment in a gigantic and misunderstood market; to drive innovation in technology and business practice; to re-appraise the essence of commerce and the people with whom we do it; all this takes decades, not years. How does this long-termism fit with a rapidly changing world? While writing this book, the solar energy sector has not stood still, the focus having shifted from lighting to solar home systems (and stumbling, as I write, towards micro-grids). Mobile phones, of course, have become smarter, consolidating their place in peoples' lives across continents. China boasts sprawls of factories where only 10 years ago the local people were subsistence farmers. The rare earth materials (such as cobalt) are now worth more than gold, and coffee is the second-most traded commodity in the whole world, second only to crude oil.

As time advances, global co-operation and understanding does the opposite. Amid the anxiety, environmental and ethical concerns seem forgotten. Emergency aid campaigns appear and disappear from our screens and newsfeeds. And yet, money still drives the world, and it motivates people as individuals and groups to feats of ingenuity, with an eye to the long-term prize – that 'absolutely enormous potential'. This is what gives me confidence that the strategy of this book can succeed, because, irrespective of reactionary cultures, politics, religion or ethics, it is going to make lots of people lots of money.

CHAPTER 27

A focus on value justifies the price

Abstract

In this chapter we try to connect the abstract 'value' offered by a fit-for-purpose, reliable energy solution, and the tangible, proven monetary 'value' required to make its success a business reality.

In summing up our journey, we provide just a few practical, real-world examples of reliable autonomous energy services, such as rural and urban electric vehicles, ICT, healthcare applications, lights and chargers, which can be implemented right here and now; but only if we can understand what is meant by 'value'.

Their success is to be judged, not by the mere provision of 'electricity' or 'energy' as a raw commodity, but for the tangible, beneficial change it brings to vast numbers of people; for the many economic, social, health, political and 'quality of life' benefits they bring.

All along, our aim has been to arrive at a realistic business model, and to point to the vast array of value propositions served by autonomous PV services. This loaded word, 'value' is the cornerstone of the strategy itself, and the entry point for anybody trying to raise capital. And it is not without its contradictions.

How do we value autonomous energy services?

We have, by and large, a one-dimensional perception of what is meant by 'value', equating it with price. This need to quantify things financially places severe restrictions upon PV-enabled autonomous energy services, whose commercial worth derives from the actual benefits it brings to everyday lives.

This is nothing to do with a shortage of applications. I could describe in detail several hundred as-yet unseen PV-enabled product and energy services ideas, ranging from toys to commercially valuable space deployments. Most of these could be built now; others anticipate improvements in existing technologies. Each of them holds commercial value. The challenge, as always, is to articulate that value in a snappy sentence to our metaphorical businessman in the 30-second allotted time window. We can understand that, by virtue of its predictability and reliability, an energy solution leads to an overall and long-term improvement in the owner's lifestyle. We can call this 'quality' and understand that it is the differentiator for our strategy. But in an economy ruled predominantly by an overwhelming *quantity* of cheap products, how do we communicate what its 'value' is? How do you put a price on a reliable internet connection for businesses, for schools and education? What is the monetary value of dependable refrigeration? For vaccine programmes involving international logistics and cool-chains from warehouses to end users? What about foodstuff and dairy for businesses and millions of households? In short, what is the 'elevator pitch' for explaining value? Let us try, then, to connect the abstract 'value' offered by a fit-for-purpose, reliable energy solution, and the tangible, proven monetary 'value' required to make its success a business reality. I will include some real-world value propositions as examples.

Economists are already putting monetary value to much of this. It might be a stretch to expect everybody to learn about the transformative power of real value over low cost in one go. (For a start, it would require the general public to change its buying habits overnight.) In the event, a compromise is likely. Knowledge sharing and grassroots improvements in local partner networks will shoulder some of the burden; but it is more important to demonstrate what autonomous energy services can actually *do*. Once people witness this, the brand can come into play to reinforce their appreciation of value.

Interestingly, the feedback I have received in writing this book has commonly taken the form of a question: Why am I sharing it – when there is significant commercial potential in this strategy (indeed, this is the whole point)? My response is a practical one. Justifying all of this in terms of value requires some detailed insight into specific sectors. By this, I mean those professionals and committed individuals who are living their vocation day-in day-out, rather than economists, NGOs or international bodies; in other words, localised, active expertise.

As an energy specialist, I am convinced of the enormous positive impact that autonomous energy services can have across the world; but, my wish is to demonstrate what PV-enabled services are capable of, and to initiate the debate for those wishing to make formal arguments in their given sector of expertise. Thus, medical professionals may define the value of dependable light, sterilisation and biomedical equipment. In education, it is teachers who can most clearly articulate the value of reliable light, computers, media players and internet access.

However, this becomes altogether more difficult when you cast wider, and try to assemble a broader critique of value from the many economic, social, health, political and 'quality of life' benefits you bring. Take agriculture as an example. When an irrigation system allows you to grow over the whole year instead of depending on the monsoon season, what are the cumulative benefits, over and above the simple matter of yield? When you apply dependable energy all along the agriculture value chain – for irrigation, drying, processing and cold storage – when do the improvements become a transformation in fortunes?

For this reason, what follows are some brief practical discussions about value for different applications of autonomous energy services. For each one, consider the principles we have explored throughout the book; that the energy services are designed holistically, are cost-optimised and incorporate an intuitive user interface with smart data collection. The energy service will be the result of a market-driven process, its design targeted for the specific service required. Visual media and a local network support these service solutions. Consider also that these fulfil our notion of a 'guaranteed service' by way of pre-failure notifications and local, capable response personnel.

ICT and healthcare

It is self-evident that the computer and digital communications have become a ubiquitous and essential part of our everyday lives. Amid our familiarity and the temptation to 'throw money at IT', it is all too easy to lose sight of what this actually *enables* at the level of individuals, households, communities and nations. In our context, it rewards us to stop and consider the specifics of value; specifically, *why* information and communications technology is essential to poverty alleviation and development.

Jeffrey Sachs, American economist and director of the Earth Institute at Columbia University, offers eight distinct contributions ICT has made to sustainable development. These are: *connectivity*; *increased division of labour*; *scale* (the ability to reach millions quickly); *replication* (the reproduction of expertise through training, for example); *accountability* (improved audit trail, monitoring and evaluation); *matching* (identifying and introducing buyers and sellers); *social networking*; and *access* to education and training (Sachs, 2008). As we have seen, providing energy enables you to switch on the devices; but there is no guarantee that you will realise the development outcomes listed above unless that energy is dependable, or at least predictable. This is the essential value offering we must make.

It rewards us to stop and consider the specifics of value; the essence of why ICT is essential to poverty alleviation and development

---

Bringing it all back home

Large numbers of very smart and talented people in the developing world live away from the rural areas of their upbringing; indeed, many live outside of their home nations. Technology helps out here, allowing distant community members to remain involved and feed back into their home societies. It is easy to transfer money by mobile phone; more importantly, they can assist with direct trade, provision of information and education, and contribute to an immediate improvement in living standards. Simply sending a text message with the market value of the crop of the village can treble the earnings of their family household for no extra effort. In addition, they can advise on family planning, preventative techniques against diseases and the political climate in relation to business and trade, for example.

This has broader ramifications. There is no longer such overt pressure to migrate, either to large cities, or further into the industrialised world. People can move back home and maintain their businesses as an accountant, writer, software engineer, editor, or designer, as long as they have a tablet or note-book PC, an internet connection (by satellite if necessary) and a source of reliable energy.

Labour migration can change, therefore, and the rich world is no longer able to fix the dice, picking and choosing the cream of the international talent pool. People who have less no longer require a visa to access more. The talent can stay at home, earn in their home nation from the wider world, and pour cash into their own local economies – not drip-feed it amid red tape or at ransom.

But again, how do we put a price on this? What is the economic value, let alone the social value, of providing robust autonomous communications to rural communities, particularly when it allows international contact with family, friends and the global marketplace? What is the true value of enabling those of poor rural origins to bring the skills and knowledge they have gained in the outside world back to their home settings?

E-commerce in landlocked, low resource countries

The need for an opening into international markets is perhaps most acute in landlocked, low resource countries, which receive little economic benefit from their immediate neighbours. Around a third of people in poverty live in these countries (most of these are in Africa). For trade, they rely on infrastructure – roads or river access to major seaports. Their best options are air and e-trade: the former has practical and economic limitations; the latter, almost unlimited potential.

Above all else, digital commerce is absolutely dependent on *reliable* electricity. It is not uncommon for the power to drop out every half an hour in many places, rendering bandwidth and speed irrelevant. These events necessitate the use of UPS systems (very fast-acting battery backup). This is a huge additional cost – as much as the computers and electrical equipment themselves – yet it is deemed an IT cost, not an energy cost. For these low resourced landlocked countries, the rural resource is the *people*, not just the land. No mains electricity grids are likely to arrive in these regions, ever; so any rural-based options for diversification from agriculture and farming would bring huge benefits. The value is clear: why not just use a very reliable, independent battery system in the first place?

> The developed world is no longer able to fix the dice. Those with less can earn in their home nation, and pour cash into their own local economies

People are starting to get the message, albeit slowly. 'Off-grid PV' is being recognised for its full life-cycle cost, but still only in terms of fuel and logistics (largely against the lifecycle costs of diesel generators). For our landlocked countries, logistics are indeed costly, but as yet the true value of *reliable* energy, and its transformative effect on national economies, is yet to be recognised.

It is something of an irony to be stuck at this stage, having to make the proposition of PV supporting ICT infrastructure – not least because, to all intents and purposes, it already does. Anyone using GPS either on their phones to location-tag a photo, or for navigation in their cars, or anyone who flies in an aircraft, habitually makes use of satellites, which are all powered by PV. Similarly, as we saw earlier, a great deal of our ICT infrastructure is already underpinned by PV technologies, while much land-based television and radio is migrating to satellite. The fact is, poor economies such as sub-Saharan Africa do not follow the historical norm. Just as communications in these regions have leapfrogged fixed line infrastructure, it is not a given that an agricultural economy has to enter a phase of heavy manufacturing, purely to ape the economic history of western nations.

So there is nothing new in this concept, technologically speaking; our strategy just extends PV that much closer to the user. These developing world networks will increasingly need to utilise non-fuel and non-grid options; and PV is by far the highest value solution.

Healthcare

Around 60% of the developing world has no access to urban-quality hospitals or medical infrastructure. Energy plays a key role in bringing the obvious benefits of medical services to these regions. Reliability is paramount here too. The simplest things have real value; in rural clinics, basic lighting enables doctors to treat patients at night. Outdoors it makes a huge difference where people have to wait, often for hours, before being seen. Energy applications for healthcare also include: high brightness inspection lights; autoclaves (sterilisers); centrifuges (for diagnosing anaemia); heart rate monitors; incubators; online record keeping and reference; digital cameras to record patient faces, injuries or condition; data storage and backup; water boiling for drinking, washing and birthing; and the vaccine cool chain – personal, health centres, hospitals, local transport, international transport and storage. There are dozens of others.

Meanwhile, cutting-edge advances in point-of-care diagnosis look set to revolutionise healthcare worldwide. In development are 'Lab-on-a-chip' technologies (LOC) – wireless devices for uploading to cloud-based solutions, notably for pandemic early detection and prevention. With one-quarter of the world without power, let alone wireless internet, autonomous energy truly shows its teeth as a valuable, life-enhancing solution.

Urbanisation and its discontents

Personally, I have always been biased against cities and in favour of the rural life. However, this is easy for me to say; in the UK, where I grew up, 'rural life' is all about green fields, organic agriculture and farming. It is an altogether different story in the developing world, where the rural life goes hand in hand with poverty, and where over a billion people live perilously close to having their livelihoods destroyed.

Urbanisation remains a huge and dangerous problem for the world. Commercial services are targeted at high population densities. Consistent mobile phone and Wi-Fi coverage, for example, acts as a magnet to millions of people on a weekly basis. As the demand for such electrical services grows, they migrate towards grid electricity. Over the last decade or so, many rural people, particularly women, have been drawn to urban businesses. Regardless of what we in the industrialised nations may think of 'foreign call centres', they have been an attractive option for those wanting a better life, for themselves and their families. But drawing millions of rural people into urban areas each day brings risks.

For many who have migrated thus, urbanisation has meant the compression of huge numbers of people into hardships even greater than those they left behind. This is no agrarian fantasy. For daily commuters and new urban residents alike, the public health hazards alone outweigh the benefits, before we even consider crime rates, or the mounting pressure on education and other public services.

What is the alternative? For a start, the equipment required for call centres is more energy-efficient, by something like a factor of 10, than it was only a decade ago. (Think of a CRT television compared with an energy-efficient flat screen; a noisy fan-cooled desktop computer with its tablet equivalent; or a 100 W incandescent light bulb against a 10 W LED version with the same light output.) This makes an autonomous PV energy solution in the region of 10 times cheaper. Meanwhile, transcription services represent a massive commercial sector in many developing nations. Both of these employment options can now easily be located in rural (meaning off-grid) areas, where most of the workers live anyway. These services will benefit from guaranteed reliability, and they can be implemented anywhere, with very low maintenance, no pollution and no fuel requirement.

But once again, how can we assess the proactive economic value of reducing daily urbanisation? Or curtailing urbanisation on a more lasting basis? Those committed to improving the lives of women, or involved with vocational learning and general education, will attest to the social benefits of rural development against the hazards of urbanisation; of which rising literacy and reduced rates of child mortality are only a start. At a micro-level, even providing sufficient light to enable learning after dusk, without incurring a fuel cost, has a massive cumulative social and economic impact.

What you are proposing to an investor is not a fixed item
with a price tag, but a process, and a key to a trend

Too often it is assumed that these improvements are mere one-way philanthropy, the happy outcome of development efforts; holding only abstract value; benefitting the immediate recipient and of no consequence to the environment, economy or society. Consequently, in the absence of a price tag, it is tempting to marginalise them as we propose the value of development solutions.

Engaging in some joined-up thinking, what would happen if we were to view them instead as *inputs* to the process? A collection of healthier, educated, upwardly mobile communities, bolstering local markets? Perhaps this is where you have the inkling of your 'pitch'. For what you are proposing to an investor is not a fixed item with a price tag, but a process, and a key to a trend. This intellectual property – the growth of self-powered communities into a mechanism for bringing confidence to an emergent market – is what is truly valuable. The prize? Being at the forefront of creating this trend, setting in motion the development cycle whose returns are vast indeed.

Selling movement

If we accept that forcing rural people and communities to live in urban areas creates more problems than it solves, we have an obvious challenge. It is essential for rural communities to trade, yet trade centres – towns on main roads, markets and rural capitals – can be tens of kilometres away. Fertilisers and cooking fuels are bulky and heavy. So is water, which is typically fetched in small volumes by women and children. Farm yields are also more valuable when sold in bulk. Conversely the cost of purchased goods can be reduced through collective buying. Meanwhile, the elderly, the sick and the pregnant – in fact anybody without access to a vehicle – are often isolated from health services.

It follows that transport is extremely valuable. But it is expensive. The cost of using livestock for transport is prohibitive: the animals require food each day, are vulnerable to disease and often too weak to work. Motorised vehicles are pricy to run (let alone to buy), and require frequent skilled maintenance, due to the harsh environment and appalling road conditions – often with poor quality fuel and components.

A containerised autonomous vehicle solution

Now consider the electric vehicle. We in modern industrialised countries baulk at their low daily range of 100–200 kilometres, although this is more than sufficient for average city or suburban use. For rural communities, to serve the local need we have identified, half of this range would easily be adequate. This is also true for top speed; for local, rural travel in the developing world, 60 km/h will be more than adequate. So the basic requirements for electric vehicles are easier to fulfil. Just as decent affordable PV lighting offsets the use of a candle or kerosene lamp (not a floodlit room), so rural electric vehicles will take the place of a cattle-drawn cart or a heavily polluting, thirsty, 20-year-old truck with maintenance issues.

In this scenario, quality – remember, I mean fitness for purpose – is everything. The type of vehicle to fulfil this will withstand the pounding road conditions, boast extreme reliability and require only basic maintenance (an inherent benefit of electric vehicles). It may well favour adaptability over luxury; so the likely candidates are not the refined domestic city car hybrids of the industrialised world markets. Rather we should envisage the long-established, rugged platforms of light commercial vehicles; the pick-up trucks and multi-purpose 'chassis cab' designs, but with batteries under the flatbed, and charged autonomously. Indeed, the electric power solution provides higher torque at low speed than their petrol or diesel counterparts – ideal for variable terrain, inclines and heavy loads. The result is a low-maintenance transport solution with a 10-year operational lifespan, zero-fuel requirement and zero emissions. The PV array will form a carport (an established practice in industrialised countries) which provides shade for the vehicle and for people. The PV can charge the vehicle directly if it is possible to leave it for a full day, say twice a week. Preferable, though more expensive, is to charge a stationary set of batteries located within the carport solution, with the charge transferred to the vehicle over a few hours, day or night.

Significantly, the entire solution – the complete stand alone carport charger system and the vehicle itself – will fit into a 20 ft shipping container, so there is a ready-made logistics infrastructure waiting to deliver them. The vehicle is ready to use from the moment the container is opened at the destination, while the carport charger can be working within a couple of hours, after which it will work for a decade with non-skilled maintenance.

In all of the above, remember that the technology is proven and widely available. The vehicle itself, yet to be built, is a matter of converting the powertrain of the ubiquitous and perennial light-commercial format. It is difficult to appreciate the full potential reach of this product, serving village communities, healthcare centres, education outreach and every shape and size of semi-rural to rural networks. The vehicles can easily support a refrigerator – either for medicines or commercial goods, and can charge dozens of mobile phones and consumer batteries each day with little impact on the capabilities of the vehicle.

What is the catch? As with most large autonomous energy solutions, the capital cost is the issue. The product does not yet exist, but taking an average figure between the popular family hybrid and a small rugged commercial van, while adding in the autonomous energy solution, I estimate a price tag of approximately $60,000. This is a lot of money, but the value proposition is irrefutable and enormous.

The value of the autonomous electric vehicle

The Millennium Development Goals (established following the United Nations Millennium Summit in 2000) identify the big five critical areas for investment and attention to alleviate poverty for rural poor: agriculture; health and education; electricity; transport and communication; and safe drinking water (United Nations, 2000). The rural electric vehicle, while not the sole universal solution, nevertheless, helps address all of these goals. It draws on multiple areas of existing funding, it is quick to implement and can be targeted to specific social and geographical areas. As such it is among the most effective starting points available.

Again, the benefits which offset the immediate capital cost are indirect, multi-faceted and sometimes subtle. Not only does this release communities from fuel and fuel-supplier dependency; it also brings trade closer to the buyer, allowing access to large regional trade locations. There are further gains. When smallholders collaborate in bulk purchasing of raw materials, seeds or fertilisers, they stand to increase the value of their produce, by avoiding the middle-man mark-up and without the burden of external market manipulation.

Benefits for education

Of these development goals, education is perhaps the trickiest to quantify. Many of the benefits of reliable energy for education are well understood, and we will mention them only briefly here: light for reading and creative learning; electricity for computers, printers, internet access, communication, speakers, projectors, outdoor lighting (especially toilets) and PA systems. Vocational learning has even higher energy demands for machinery, craft-work, e-commerce and so on.

Not least of the challenges for education is *attendance* – actually having the children to teach in the first place. The primary-age pupils who do make the walk (often a few kilometres) are invariably physically and mentally tired once they arrive. Secondary schools, meanwhile, are remoter still, and far less common. The barriers to education occur at the most critical time of a young person's development – just when they become valuable to the everyday necessities of the family. A school bus, as for anywhere in the world, guarantees the basic value of school attendance; the difference being that it is free to run and free to use. And when I say bus, I mean the broad sense of our multi-purpose chassis-cab or flatbed, converted with seats or cushions in the back (a common sight in the developing world).

Meals for school children and staff could be prepared by the communities (by parents, perhaps, in lieu of tuition fees) and driven to school for lunch-time, thus benefiting from economies of scale.

Fuel, clean water, school meals and power

Consider one of the prime reasons children fail to attend school: children and women spend several hours every day fetching water and firewood. Typically, this is because they can only carry so much at a time (though often these loads are impressive to the point of physical injury) and the sources are kilo-metres away. Now extend the reach and the value of the very same vehicle we have mentioned, with a view to public health. Water-borne diseases account for maybe a quarter of child deaths. Our 'school bus' is perfectly capable of collecting full loads of water for households while the children are at school, making multiple trips to cleaner and more hygienic sources which may otherwise have been denied those on foot. Fuels for cooking, in the absence of environmentally sound alternatives in the interim, can also be sourced in bulk using this transport, from trade centres which avoid the mark-up of semi-rural to rural suppliers. (Kerosene and firewood can be twice the cost for the people who need them most, by virtue of distance.)

Further applications

So far, our single transport solution demonstrates that a capital cost brings a return on investment – a 'value' – which defies traditional methods of measurement. This is because we expect a direct outlay to map to a direct outcome; a profit out-of-the-box, as it were. But as we are not used to working within a poor region, or a market with everything to gain, strictly speaking these rules do not apply. The genuine reward from our single product is not a hit at the point of sale. It is indirect. It is multi-faceted, affecting not only the customer at the point of use, but the culture and the society in which that customer lives, works and raises a family.

Returning to the example, then. In addition to a delivery platform, the rural electric vehicle is, in effect, a mobile power station, comparable in capacity to a diesel generator used for a modest city home. It can power many of the devices essential for ICT, including satellite or GSM equipment, making it a fully mobile communications centre. It is also, with modest peripheral conversion, a water pump, using either the charge station's energy system or the vehicle's own generating capacity.

As food for thought, and broadening the concept, consider this solution in the form of a tractor. Or for heavy lifting. Forklift trucks, for example, are already widely used in electric form. Think also of the versatility of electric scooters and bicycles. What about electric mobility aids, such as wheelchairs?

Earlier, in Part Five, we interpreted the global market in terms of four sectors: the industrialised world; disaster and emergency response; small island states; and the developing world. We have looked at electric trucks in remote and poor regions above. Now think of our village vehicle in a small island state, where fuel is expensive, skilled maintenance limited, and where spare parts are expensive and difficult to source.

Meanwhile, applications for disaster and emergency response are prime candidates for PV solutions, by virtue of their need for rapid, remote and autonomous deployment. Search and rescue equipment, such as remote-controlled robots, thermal imaging cameras, acoustic sensors, and air and gas monitoring equipment (as used in volcanic regions, for example). All can be served by our autonomous electric solutions. This requires vision – the boldness to immerse fully in these markets; we know that it would cost too much to develop any single such product on its own. But a strategy which acknowledges the breadth of solutions, and designs accordingly for versatility, may yet make this concept of value a reality in the marketplace.

Finally, think of urban uses for these vehicles. Anyone who has driven squinting through any noisy major city in the developing world, and blown their nose at the end of the day, will know of pollution. We have a vehicle with zero running cost and zero emissions that places no energy burden on the mains network. In fact, it could even support that network – electric vehicles can be used for balancing peak demand on the utility grid.

All well and good, but hang on a minute. We began our journey looking at the shaky premise of grid infrastructure; the decades of broken promises to provide mains electricity to rural locations; the blind assumption of the West that industrialised grids provide the only answer to the world's energy need. Throughout, we have searched for a means to realise a degree of self-determination for remote peoples, rather than prolong their fuel dependency.

If there is real value in this impressive litany of PV-electric devices and products shown above, it is precisely their ability to sidestep the infrastructure of electricity grids. Compared against a traditional infrastructure project, this electric vehicle solution is extremely cost effective and facilitates a whole host of fundamental rural poverty solutions. And it is not difficult to implement. Critically, there are multiple areas of revenue generation for the supplier and their local partner. For example, many of the vehicle value propositions can draw affordable amounts from lots of local people. Over the life of the vehicle, this income is significant. In fact, local businesses can be enabled by the capabilities of the vehicle.

Here, then, is our value. Not a number on a balance sheet, nor a short-term return on investment, nor yet a quick buck; but the realisation of a strategy that can energise a giant nascent market to power, from static impoverishment to profitability, via social inclusion and mobility, improving public health and quality of life. Changing the world, it would seem, is good for business.

A force for democracy

Poverty is synonymous with remote, rural areas. Aid organisations and governments alike target their grants, loans and investments at these communities accordingly. Yet these initiatives are the most difficult to monitor or assess. It is vital to understand and communicate the movement of money and assistance. Transparency and accountability are a prized arsenal in a war of information against poverty; sometimes, establishing simple communications with the recipients is all it takes to improve the effectiveness of finance.

Here again we see the expanded scope of digital communications. Governments, for example, can release data on fund allocations on a regular basis, as well as when budget spending takes place. They can also directly send information to the rural recipients – the schools, medical centres and individual households. With simple phones or internet connectivity, those recipient communities have the technical ability to confirm whether this money is making its way to the intended destinations.

This grass-roots transparency is powerful. If you provide reliable communications to a village-sized community, you are creating a dialogue, in which the information coming out of the village is equal in importance to that which flows in. This assists in increasing the effectiveness of aid budgeting, makes the most of national resources, while all the time reporting on the success of the programme.

Distributed solar electric generation and storage offers the opposite of centralised power, replacing it with generation closer to the point of use, and taking control out of the hands of national regimes

---

You will see by now that there is more going on here than a change of power source. The 'value' of our strategy goes far deeper than meeting an immediate need. When we talked of grids and fuel dependency earlier, we dwelt upon the social or psychological dependency as much as the physical need for resources. But fuel dependency inevitably has a political bearing in addition.

'There are 22 countries in the world that earn 60 percent or more of their national income selling oil. All 22 of those are dictatorships or autocratic kingdoms.'

PV undercuts the ability of strongmen to buttress their power through the control of the electricity grid, which was the case with the late dictator of Zaire, Mobutu Sese Seko. 'Mobutu cut off villages, if they wouldn't back him.

The other great example is Lenin's line that the dictatorship of the proletariat is the rule of the Soviets plus electrification' (Warren, 2012).

Former CIA Director James Woolsey talked of the potential of PV to reduce conflict and suffering in oil-rich nations.

Distributed solar electric generation and storage offers the opposite of centralised power, replacing it with generation closer to the point of use, and taking control out of the hands of national regimes, totalitarian or otherwise. We can talk of quantitative change – that PV is cheaper than extending the grid, or cleaner than kerosene. But only if we are prepared to acknowledge that ours is a completely different way of understanding the role of energy in people's lives. Its long-term value is no less than a radical shift in outlook, with social and political repercussions. 'Personalised Energy' services can be a force for democracy and freedom; affording individuals, families and communities the self-determination to conduct their lives, the materials and media with which to pursue political freedom from oppression and profit from the freedom of markets – a power structure in their own right, but one in which people, in partnership, can exert a degree of local control.

A new way of creating sustainable, reliable power
is a formidable force for progress in the world

Absolutely enormous potential

In summing up our journey, I have provided just a few practical, real-world examples of reliable autonomous energy services, such as rural and urban electric vehicles, ICT, healthcare applications, lights and chargers, which can be implemented right here and now; but only if we can understand what is meant by 'value'. Their success is to be judged, not by the mere provision of 'electricity' or 'energy' as a raw commodity, but for the tangible, beneficial change it brings to vast numbers of people.

To 'sell daylight' is to reimagine value for society. When we talk of value in relation to commerce or funding, we go in search of a price tag – an expression of our 'return on investment'. Indeed, we can place monetary figures against some of the initiatives and practices of development, and this is necessary to get at least an arbitrary handle on scale. But we can only go so far attaching mathematical models against the scale or diversity of giant continents, where reliable, guaranteed services are truly transformative, essential rather than desirable.

But we then start to see that this transformation is another way of looking at 'value'. A new way of creating sustainable, reliable power as a formidable force for progress in the world. Where we have undertaken to involve local supply lines and support networks, we have unleashed the power and expertise of local populations. And by virtue of the technology itself – autonomous, reliable, clean, locally maintained – we have introduced a degree of self-determination to remote communities otherwise dependent on fuel, grids and sometimes corrupt monopolies or regimes.

It is this feedback loop, this virtuous circle, which seeds the vast, dormant, remote, poor areas of the world into rich plains, pregnant with cultural and economic progress. When families, communities and entire regions are able to access local healthcare, education and transport, the improvements to their lives defy a price tag description. And yet, it is this life-changing 'value' which provides the key, the spur to invest in an upwardly-mobile, rising market culture.

Before us, then, lies an immense capital opportunity, in the emerging markets of the developing world, small island states, disaster and emergency response, and industrialised nations.

Everything I have described in these pages is entirely possible

---

Nobody is claiming that this strategy is easy. The methods for identifying these markets might not be readily obvious, because they differ from more familiar, traditional market sectors. This is partly due to the way we are used to doing things and our ingrained ways of thinking. The highest value solutions for emerging markets may hold little meaning for the industrialised world. This should be no surprise, given what we know of utility grid systems – which meet all the electrical needs of industrialised nations but are ridiculously inadequate against the scale of the developing world.

And yet, everything I have described in these pages is entirely possible. It requires no new technology, no extra money beyond existing energy-related spending, and no specialist skills or equipment beyond what is easily accessible. The absolutely enormous commercial potential for global sales of autonomous energy services is within our grasp. There it is, above you.

CHAPTER 28

A civil dawn: life with inclusive autonomous energy services

Abstract

This final chapter provides a positive narrative of just what a future network of autonomous and inclusive energy services could look like for a developing nation.

We narrate a typical day for a drone operator as they observe a multitude of autonomous, PV-enabled equipment that combines to describe a nation and its people in socio-economic progress.

In keeping with the central principles of our strategy, we make reference to many of the funding mechanisms that can make our vision a reality.

Most importantly while reading this, remember that everything that is described can be implemented now, within existing energy-related expenditure, using widely accessible technology and easily accessible skills.

It is an hour before first light. A city mimics the behaviour of its inhabitants as they prepare for the day; its discrete, isolated infrastructure stirs and transitions to the rhythm of the sun.

Silently, diligently and largely unnoticed, batteries everywhere begin to receive electricity from above. Along the road, streetlights extinguish in sequence, as if someone were walking west and switching them off manually. Cat's eyes stop flashing and the billboard lighting takes a rest. The city's commuters start to amass, preoccupied with their everyday business, and the morning's traffic begins to build.

A drone flies overhead. It is equipped with a camera for traffic reporting, and, like the rest of its fleet, it is charged from solar PV. From up here, its operators can be in no doubt of the changing society it surveys. The drone navigates the city streets, aided visually by the numerous PV panels – clean and flat rectangles with shiny edges against a haphazard urban backdrop. All face the equator, and every now and then they throw a reflecting glare as the sun continues to rise.

Thousands of PV panels like these are now mounted throughout the country. The major roads are lined with rows of single panels atop streetlights, as if lining a grand driveway. Panels above traffic lights look like orderly groups of sentries. On bus stops they are out of sight to the general public, while those on parking meters and road works are so common they no longer merit a second glance. Rubbish bins now signal when they are full.

http://dx.doi.org/10.3362/9781788530705.028

Uptown, the cafés automatically extend their sunshades to capture the sun's energy before it heats up the room, ready to provide USB charging and to light up the shop front the coming night. Arrays sit on containerised stores opening for business, in the parks for the fountain displays, and in less manicured areas, supporting the water supply. Arrays grace rooftops of every type, from the city centre to the suburbs. All sit bathing in the new day sun.

It is morning. Odilla arrives for work before the day really heats up, but she is still sweating. She wears a company-branded baseball cap and a smart short-sleeved shirt, and is relieved to walk into her control room. This is little more than a small cabin in the suburbs of the city, but it has cool air, internet access and a charging socket. It houses a drone control desk. Odilla plugs her phone in, switches on the console and makes herself comfortable as the system boots up, then clicks 'Initiate Drone'. A greyish picture appears on the screen for a few seconds, before giving way to a blurry bright hue. The image focuses in and out, pans left and right, then vertically. She is now looking at the outside of her control hut on the screen.

Evie, a young and conscientious man wearing a baseball cap, appears and visually checks the drone. He leans in front of the camera and gives a thumbs-up with an enthusiastic smile. He hopes his diligence will make him a controller one day. On Odilla's instruction, the drone auto-starts and lifts off the ground. She pushes the joystick forward briefly and the drone nods to her colleague outside. He tips his cap in response.

As the drone rises, the roof of the cabin appears on screen. The aluminium frames of the roof-mounted PV modules gleam in the morning sunlight. In a calibration procedure, the camera auto-focuses on the dock's display, which shows 40%, and charging from its dedicated array. Odilla instinctively checks the drone itself, which has 99% charge, before selecting the day's inspection route.

As the drone leaves the station, its itinerary is shown on Odilla's map screen as a scattering of dots; mostly green, with a few orange (which is OK this soon after dawn), and on this particular morning, a few red. Two bus stops in busy commuter areas are not functioning. For the first, the on-screen data reveals the shelter passenger information display was switched off at 3 a.m. due to low battery power, and the lights went out at 5 a.m. The drone arrives and focuses on the shelter until the picture becomes sharp. There are leaves covering parts of the PV modules on the roof. Odilla logs the observation, though the shelter should have signalled this issue itself. The second shelter is more easily diagnosed – bird mess on the PV. Odilla knows this area of town and is familiar with the offending creatures: intimidating vulture-like things that look like they're from prehistoric times. They roost in the tall trees beside the busy road and even on structures above the carriageways. (Inexplicably, they are supposed to be good for tourism, although it's the tourist drivers who are so regularly distracted by them.)

A bus arrives at the stop, represented on her monitor by yellow dots, moving along the line of green asset indicators like Pac-Man characters. (Other drones and charging stations appear in blue.) PV-enabled location systems were fitted to all buses in a single month as part of a public transport policy initiative a while back. But it was the on-board Wi-Fi that has seen bus operator revenues grow steadily ever since.

Odilla's augmented display forms a mosaic of the basic city layout – particularly, the electricity network. Drones just like hers inspect hundreds of kilometres of transmission cables far and wide. The power systems and hardware generally go unnoticed – outshone by the illuminated and glamorous advertisements that pay for the capital's very own 'Internet of Things'. Meanwhile outside of the city, there are further networks used for conservation, animal welfare, environmental protection and various development programmes. These are paid for by a different kind of commercial service. Here, the 'curse of natural resources' has finally become a purse. Local companies now quantify mining extraction, and monitor pipelines for the government. This is big business. These local enterprises, once outraged by the exploitation and lost revenue of the past, are now squarely in command of the numbers, emboldened by reliable phones and internet access.

Odilla's drone continues along the route of a large pipeline as part of its daily task. These days people gather just as the day's heat starts to build. They have come for the free online access at one of the pipeline control stations. This is the brainchild of a local group who maintain the PV system, who figured that since the batteries are nearly always charged by midday, they could keep the place busy and keep vandalism at bay. Where once there was barbed wire, cameras and security guards, now here are groups of locals surfing the internet. Some spend the rest of the day watching videos or sport, while others check the news and their email. A social gathering in the strangest of places – standing next to a cathodic protection controller, video chatting as if they were in their own homes. Odilla locates a nearby docking station at a telecoms tower, and instructs the drone to auto-land. An hour's lunch break for the controllers, a fast charge for the drone, and a data upload: one PV system chatting to its counterpart in orbit.

After lunch Odilla's drone flies above the nature reserve. This is her favourite part of the week, but most locals don't have much time for the 'countryside' (as mzungus, or white people, call it). 'Everything wants to hurt you' her friend Juvenal says. 'Even the plants!' On her screen, solar-powered animal tracker tags show up as green dots. The accompanying code tells Odilla they are a herd of elephants, but she zooms the camera in on them anyway and flies at a distance she knows will not unsettle them. The drone automatically uploads data from their tags. Several other green items confirm that all is operating healthily: electric fencing, anti-poaching security cameras, night observation cameras, emergency outposts (each with communication, light and water), and the security huts at the park entrance. These are identical to the drone cabin, each with their PV systems, cooling vents and communications aerials.

Odilla flies towards one of these huts in the park. Her childhood friend emerges in ranger uniform and tilts his hat good afternoon. She 'nods' an elegant return salutation before turning to fly up-country. Their walk to school used to take them 2 hours. Now she flies it in 15 minutes. From this high up, she no longer needs to navigate using the screen. PV arrays shine like silver against a backdrop of red earth, brown foliage, and sometimes sand.

One of these arrays, an orderly rectangle among the randomness, sits near a watering hole. Here, standard and infrared cameras keep watch, capturing stunning images and information about the habits of the visiting wildlife. Dozens of animals are busy drinking as the drone flies high overhead and collects its data, but the most impressive creatures (and the most sought-after) won't show up for quite a few hours yet. Further on, and protected from the animals who would otherwise devour them is a pristine and isolated square of green crops. It's not all prevention. A keen eye will make out animals grazing underneath a PV array – a canny initiative by the locals who knew which plants would grow in the shade, and only the shade.

Flying now over a rural village of brown dwellings, each notably sporting a small panel on the ground outside. The women take them out with the day's washing and bring them in together. A teenager tends a dozen of them, charging what look like portable lights; remembering to point the panels towards the sun, with the devices in the shade behind.

A small orderly queue leads to a nondescript PV-panelled building. The bank is open. A similar number loiter in front of a kiosk; more still sit shaded by the healthcare building. A lone vehicle kicks up dust and Odilla instinctively looks for its container charger. At the edge of the village some men are sat under a tree, discernible by the panels scattered nearby and the cables extending into the shade. A larger building, most likely a school house, boasts 20 or so panels in an orderly arrangement.

Dusk. The blazing aggression of the afternoon now yields to a diffuse twilight. Between day and darkness, the hidden world of autonomous energy springs visibly to life: discrete, unaided and self-determined. Home security lights blink, and garden markers illuminate as if from nowhere. Potholes and other road hazards announce themselves as flashes of orange, while traffic management services get on with their jobs.

Odilla clicks 'Land at base' on the screen and watches the drone's graceful descent towards a small fenced compound. Guidance markers appear on the screen above the circled 'D' marked on the ground. The drone lands and as the propellers come to a stop, the screen reads 'Data download complete', followed by 'Recharging'. A figure appears in the background of the camera picture. The light from his baseball cap bounces as he walks. He kneels down to look into the camera, tips his hat, winks and smiles, then gives a thumbs-up to his colleague.

A couple of hours later, and it is night. A bus shelter casts light into the road, its information display adding a touch of colour to the proceedings. At this busy junction, the traffic lights have brought a measure of order – when they're obeyed – and the amenity lighting has made the place noticeably less perilous.

But the pedestrians steal the show. A snaking line of white cottons, bright greens, yellows and blues, their suits, dresses and headgear are all lit up from the dancing glare of their torches, phones and reflectors, marking the evening exodus from the city. Even some of the bags illuminate when opened. Bicycles look like fireflies as they weave along the roads, bouncing violently every now and then. These days, the passing vehicles notice the cyclists and pedestrians in advance and afford them some room, sometimes assisted by the flashing orange hazard lights. But they still churn the dust into a frenzy – more noticeably than ever now it is a psychedelic wake lit up from every angle.

Lining the road for a couple of hundred metres on both sides, are the warm glow from houses, cool white from local stores and kiosks, and the unmistakable blue light of televisions. The choreographed reactions of a crowd outside a local bar reveal that they are watching a football match. Condensation on their drink bottles reflects the surrounding LED advertisement. Further into the hills, the TVs in the kiosks attract their niche audiences for news, soap operas and music videos. The shadows of the standing viewers dance with the programme at the edge of the darkness.

The village is finally quiet, but it never gets completely dark. Being the hottest season, many people are studying or working into the night, when it is coolest. We all remember the flickering naked flames, and the smell. It wasn't that long ago. It still seems magical to have light from the sun to use at night, even for those of us who understand how the solar lights work. Most people we know are not too interested in what is actually going on, they just smile saying that they have 'captured some sun!' These phrases have carried well. The drone operators say 'we fly using light energy, so no matter how fast, it is always light speed!' Others say that they are 'owning some sun'. But they all say something bigger: sunlight in a gift box, sold to whoever is happy to pay. At the Kweri kiosk they're selling more beers than ever. However you want to describe it, as long as the sun rises, tomorrow will be a good day for business.

References

Acemoglu, D., & Robinson., J. A. (2013). *Why Nations Fail: The Origins of Power, Prosperity and Poverty*. London: Profile.

Associated Press. (2012, July 31). *Electricity grids fail across half of India*. Retrieved January 19, 2017, from http://www.hurriyetdailynews.com/electricity-grids-fail-across-half-of-india-in-second-day.aspx?pageID=517&nID=26782&NewsCatID=356

Bloomberg. (2012, June 21). End the Subsidies of Fossil Fuel. *Bloomberg Articles*. Retrieved from https://www.bloomberg.com/view/articles/2012-06-21/end-the-subsidies-of-fossil-fuel

Cool Materials. (n.d.). 24 Ridiculously expensive everyday items. Retrieved from http://coolmaterial.com: http://coolmaterial.com/cool-list/24-ridiculously-expensive-everyday-items/

Demirguc-Kunt, A., & Klapper, L. (2012). *Measuring financial inclusion: the Global Findex Database*. Washington DC: World Bank. Retrieved from http://documents.worldbank.org/curated/en/453121468331738740/Measuring-financial-inclusion-the-Global-Findex-Database

Diamond, J. (2012, June 7). 'What Makes Countries Rich or Poor?' Review of Why Nations Fail: The Origins of Power, Prosperity, and Poverty, by Daron Acemoglu and James A. Robinson. *New York Review of Books*. Retrieved from http://www.nybooks.com/articles/2012/06/07/what-makes-countries-rich-or-poor/

Grameen Bank. (2015, December). Introduction. Retrieved December 20, 2016, from Grameen Bank: http://www.grameen.com/introduction/

GSMA Intelligence. (2015). *The Mobile Economy: Arab States 2015*. Retrieved February 12, 2017, from http://www.gsma.com/publicpolicy/wp-content/uploads/2013/01/gsma_ssamo_full_web_11_12-1.pdf

Hammond, A., Kramer, W. J., Tran, J., Katz, R., & Walker, C. (2007). *The Next 4 Billion: Market Size and Business Strategy at the Base of the Pyramid*. Washington: World Resources Institute.

Hardoon, D. (2017). An Economy for the 99%: It's time to build a human economy that benefits everyone, not just the privileged few. Oxfam. doi:10.21201/2017.8616

International Energy Agency. (2011). *Solar Enery Perspectives*. Paris: IEA Publications. Retrieved from http://www.iea.org/publications/freepublications/publication/Solar_Energy_Perspectives2011.pdf

International Finance Corporation. (2012). *Lighting Africa Market Trends Report 2012: Overview of the Off-Grid Lighting Market in Africa*. World Bank. Retrieved December 16, 2016, from http://www.dalberg.com/documents/Lighting_Africa_Market_Trends_Report_2012.pdf

Krause, K. (2010). *The True Size of Africa*. Retrieved from Kai Krause: http://kai.sub.blue/en/africa.html

http://dx.doi.org/10.3362/9781788530705.029

Leach, A. (2014, January 29). Rwanda's next education challenge: teacher training and employability. *The Guardian*. Retrieved from https://www.theguardian.com/global-development-professionals-network/2014/jan/29/rwanda-education-teacher-training-youth-unemployment

Lucintel. (2012). *Growth Opportunities in the Global Beauty Care Products Industry*. Texas: Lucintel. Retrieved from http://www.lucintel.com/beauty_care_market_2017.aspx

Mohn, T. (2013, October 7). *Travel Boom: Young Tourists Spent $217 Billion Last Year, More Growth Than Any Other Group*. Retrieved from Forbes.com: https://www.wysetc.org/2013/10/07/forbes-the-new-young-traveler-boom/

Moyo, D. (2009). *Dead Aid: Why Aid Is Not Working and How There Is Another Way for Africa*. London: Penguin Books.

Sachs, J. D. (2005). *The End of Poverty - Economic Possibilities for Our Time*. New York: Penguin Press.

Sachs, J. D. (2008). *Common Wealth: Economics for a Crowded Planet*. New York: Penguin Press.

The American Physical Society. (2009, April). This Month in Physics History – April 25, 1954: Bell Labs Demonstrates the First Practical Silicon Solar Cell. (A. Chodos, Ed.) *APS News, 18*(4). Retrieved from APS News Archives: https://www.aps.org/publications/apsnews/200904/physicshistory.cfm

United Nations. (2000). *United Nations Millennium Development Goals: Background*. Retrieved March 30, 2017, from United Nations Millennium Development Goals website: http://www.un.org/millenniumgoals/bkgd.shtml

United Nations. (2014). *Population and Development in SIDS 2014*. Retrieved February 17, 2017, from UN Sustainable Development Knowledge Platform: https://sustainabledevelopment.un.org/content/documents/1960Population%20and%20Development%20wallchart.pdf

United Nations Development Programme. (2011). *Fast Facts: Universal Energy Access*. New York: United Nations. Retrieved December 16, 2016, from http://www.worldenergy.org/wp-content/uploads/2012/10/PUB_World-Energy-Insight_2011_WEC.pdf

United Nations World Commission on Environment and Development. (1987). *Report of the World Commission on Environment and Development: Our Common Future*. United Nations. Retrieved from www.un-documents.net/our-common-future.pdf

Warren, C. (2012, June). Solar security: Former CIA director James Woolsey has a recipe for US security: feed-in tariffs and lots more PV. *PHOTON International, 2012*(6), p. 34.

Warschauer, M., & Ames, M. (2010). Can One Laptop Per Child Save The World's Poor? *Journal of International Affairs, 64*(1), 33-51. Retrieved from http://www.jstor.org/stable/24385184

World Energy Council. (2011). *World Energy Insight 2011*. London: First. Retrieved December 16, 2015, from http://www.worldenergy.org/wp-content/uploads/2012/10/PUB_World-Energy-Insight_2011_WEC.pdf

Index

'green' energy, 27, 59, 70, 89, 99, 138, 230
'internet of things', 187, 295, 317

absolutely enormous potential, 187, 280, 299, 313
Abu Dhabi, 78
academia, academics, 189
Addis Ababa, 89
affluence, 10, 15, 26, 150, 154, 177, 237, 243
Africa, 15, 24, 117, 145, 149, 173, 251, 280
agriculture, 149, 150–51, 219, 224, 302, 304, 305, 308, 318
aid, 146, 148, 177, 224–5, 296, 299, 311
AIDS, 152
alternating current (AC), 58, 61, 65, 106, 116, 136, 280
altruism, 177, 187, 295
apathy, 11, 16, 172, 177
appliances, 13, 59, 86, 101, 108, 112–13, 123, 251, 278
automobiles. *See* cars
autonomous PV systems, 61, 81, 87, 90, 97, 102, 105–6, 137, 306

backup, 35, 60, 70, 84, 138, 231, 233, 251, 259, 304–5
bags (solar), 91, 137, 218
bank, banking, 160, 163, 166, 200–201, 237, 286
Bankside Power Station, 34
base-load generation, 35
batteries, 53, 128
    battery life, 88, 113, 187, 209, 247
    capacity, 106, 115, 129, 247, 251
    car, vehicle, 9, 98, 308
    discharge, 109, 115, 247
    disposable, 96, 125, 201, 203, 233
    lead acid, 29, 114, 128, 247
    lithium, 185, 251
    performance, 206, 250, 256
Battersea Power Station, 34
battery chargers, 55, 63, 69, 80, 140, 205, 262
beauty, personal care market, 153
bespoke systems, 64, 83, 117
bias
    human, cognitive, 73, 176, 179, 294
    statistical, analytical, 148, 230
bidding, 115, 127, 148
biomass, 9, 18, 170, 267

BOS (balance of systems), 127
brand, 100, 135, 198, 202, 217, 236, 250, 281
    brand loyalty, 42, 204, 256
    brand value, 42, 54, 123, 139, 191, 249, 252, 255, 282, 285
    global brand, 75, 135, 141
budgets, 121, 126, 189, 213, 267, 311
building integrated PV (BIPV), 60, 66
buoys, 82, 87–8
bureaucracy, 128–9, 204, 294, 296
bus shelters, 62, 64, 99, 318

calculators, 62–3, 82, 90, 99, 206
capital, 139, 167, 279, 299, 301
    capital cost, 59, 66, 161, 191, 263, 270, 281, 284
    capital expenditure, 159–60, 199, 217, 226, 244, 262
    capital funding, 28, 139, 160, 280
capitalism, 188, 294
cars, 13, 51, 98, 191, 210, 307
    driverless cars, 187, 295
cathodic protection, 84, 317
charcoal, 9, 18, 231
charge controller, 61, 109, 113, 115, 260
charge indicator, 49, 109
charging
    lights and chargers, 95, 99, 301, 313
    mobile phone, 29, 67, 77, 95, 262, 275
charity, 147, 152, 187, 225
cheap products, 97, 101, 130, 179, 181, 189, 262, 301
children, 9, 18, 150, 155, 161, 163, 257, 307, 309
China, xix, 13, 15, 79, 119, 138–9, 155, 184, 299
clean coal, 41
climate, 64, 117, 139, 150, 173, 185, 248
    climate change, 37, 19, 147, 221, 235, 296, 298
coal, 12, 34, 36, 41, 60, 100, 176, 200
coercion, 153
collateral, 166
colonialism, 258
colour (skin), 73, 175
commerce, 68, 83, 163, 204, 234, 267, 299, 313
    commercial landscape, 204, 217, 223, 234
    commercial model, 270, 282, 298
    commercial momentum, 214

commerce (*cont.*)
  commercial organisations, xix, 78, 180,
    188, 264, 280
  commercial potential, 10, 138, 187, 192,
    195, 212, 243, 258, 278, 285, 313–14
  commercial strategy, 230, 232
commercial PV, 60
common core technologies, 87, 205–6, 213
communication theory, 72, 295
compliance, 62, 127, 224
computers, 22, 106, 111, 131, 165, 249,
  257, 302
confidence, 198, 226
connectivity, 232, 298, 303, 311
consumer choice, 161, 189, 236
consumerism, 156, 182, 189, 256, 281
consumer products, 55, 63, 82, 97, 120, 184,
  186, 205, 244, 247, 272, 281
consumers, 15, 23, 74, 100, 135, 190,
  236, 266
contamination, 69, 170
control systems, 35, 85, 207, 211, 271
cooking, 7, 14, 18, 163, 169, 231
corrosion, 83, 88, 128, 132
corruption, 113, 149–50, 169, 177, 229
cost
  cost analysis, 85
  low cost, 100, 188, 223, 262, 267,
    295, 302
  operational cost, 95, 101, 245, 283
  optimised cost, 102, 258, 302
  purchase cost, 13, 95, 101, 160, 189,
    204, 258
crops, 150, 212, 219, 271, 318
culture, 10, 146, 151, 160, 264
  assumptions, errors, 72, 257, 274
  consumer culture, 156, 281
  cultural change, 32, 155
  entitlement, 138
  entrenched, 37, 160, 177, 204
  growth, 278, 313
  neutrality, 47, 54, 72
  patriarchal, 73, 274
  popular culture, 50
  symbolism, 74
customs (importing goods), 211, 223, 268
cycle life, 247

daily depth of discharge (DDOD), 247
data
  data collection/data logging, 84, 105–6,
    113, 127, 208, 303
  data storage, 305
  market data, 282
  misleading, xviii, 151, 201, 230
  phone data, 52, 252
  usage data, 270
death, dying, 18, 29, 74, 146, 155
demand (economic), 23, 191, 197, 203, 231,
  237, 259, 282

demand (for energy), 6, 7, 14, 15, 23, 28, 60,
  111, 311
democracy, 41, 293, 311
Democratic Republic of Congo, 176
denial, 155, 179, 291
dependability, 5, 19, 21, 26, 100, 118, 170,
  235, 243, 301
dependency (economic), xiii, xxiii, 16, 28,
  150, 291
design
  hardware, 105
  holistic, 196, 295, 297
  integrated, 108, 127, 199
  specification, 113, 258, 261, 282
developing world
  economies, 152, 160, 235
  energy services, 37, 87, 179, 185, 231,
    255, 258
  fuel use, 9, 15
  grid electricity, 5, 21, 23, 28
  harming, 20, 27, 294
  innovation, 279
  market, 65, 214, 217, 222, 230, 283, 313
  products, 65, 210, 282
  resources, 150, 176
  technologies, 165, 256
development, 278
  goals, efforts, 96, 293, 303, 306
  international development sector, 148,
    189, 224, 297
  money, 152, 159, 177, 179–80, 198
  role of energy services, 163, 196, 278
  sustainable development, 182, 220
DFID (Department for International
  Development), 224
DFMA (Design for Manufacture and
  Assembly), 210
diesel, 9, 29, 100, 201, 231
direct current (DC), 58–9, 61, 89, 106, 116,
  129, 136, 284
disaster and emergency response, 147, 208,
  212, 218, 226, 310
disease, 18, 29, 89, 150–52, 197, 219, 303,
  307, 309
distributed generation, 23, 60
distribution networks, 39, 101, 182,
  269, 272
diversification, 202, 205, 270
diversity, 39, 53, 169, 181, 292
domestic PV, 60
donations, 139, 153, 156, 179, 236
drone, 315
dynamic commercial capability, 217, 227

earning, wages, xix, 11, 59, 139, 153,
  201, 303
e-commerce, 17, 186, 304, 309
economists, xviii, xx, 148
education, 17, 112, 150, 163, 303, 309
ego, 180, 295

emissions, 11, 19, 21, 58, 99, 292, 298, 308
energy access, 6, 20, 100, 170, 263
energy consumption, 11, 23, 29, 48
energy efficiency, 13, 29, 106, 280,
     283, 306
energy ownership, 112, 116
energy poverty, 7, 146, 162, 177, 179, 188,
     296, 297
energy services, 8, 32, 50, 67, 68, 147,
     234, 301
environment, 16, 19, 29, 182, 297
ergonomics, 53
ethics, 178, 180, 223, 295
Ethiopia, 259, 262
Europe, 34, 60, 74, 123, 248

family planning, 163, 303
farming, 128, 271, 304, 307
finance, 159, 220, 311
     access to, 159
     microfinance, 166, 223, 235
flexible PV, 207, 262
food, foodstuffs, xix, 12, 51, 68, 96, 149,
     220, 280, 302
frequency (of electricity), 15, 35, 59–60, 136
fuel, 3, 8, 11, 51, 82, 230, 309
     disconnection from, 11, 16
     fossil fuel, 8, 83, 298
     fuel dependency, 15, 27, 29, 66, 147,
          165, 309
     offsetting, 3, 10, 64, 69, 97, 164
     price of, 19, 220, 293
fuel cells, 231
funding, 37, 126, 128, 139, 200, 224,
     234, 308

gas, 4, 12, 17, 68, 83, 117
gender, 73, 150, 166, 274, 306, 309
generation capacity, 7, 23
geothermal, 28, 176
global markets, 213, 217
governments, 17, 24, 131, 204, 236,
     294, 311
GPS, 69, 293, 305
Grameen Bank, 166
grants. See subsidies
grid electricity, 3, 230
     alternatives to, 31, 297
     central generation, 12, 24, 27–8
     interconnected grids, 33
     limited mains access, 6, 231
     people without access to, 7, 38, 231–2
     reliable networked grids, 5, 38, 197, 231
     shortcomings, 28
     smart grids, 23
     spurs, 7, 24–5, 200, 232, 280
     unreliable networked grids, 6, 231, 243–4
growth, 14, 96, 135, 180, 196, 214, 245,
     255, 257, 277, 285
GRP (glass reinforced plastic), 128

guaranteed service, 123, 236, 280, 303, 313
guaranteed services, 243

handshaking, 251, 284
hardware, 249
hardware design. See design: hardware
healthcare, 68, 153, 282, 305
heating, 12, 169, 234
High Intensity Discharge (HID) lamps, 88
holistic solutions, 106, 196, 199, 295, 297
hospitals, 87, 134, 305
human behaviour, 91, 155, 174, 295
humidity, 89, 206, 221, 256–7, 283
hybrid systems, 297
hydropower, 33, 35, 40, 267

icons, 80, 123
ICT (information and communications
     technology), 67, 303
India, 15, 21, 74, 138, 166, 233, 262
industrial applications, 65, 87, 184
industrialised world, 3, 12, 65, 203, 213,
     218, 256, 273
Industrial Revolution, 8
infographics, xviii, 74
innovation, 156, 186, 279, 292, 299
institutions, 150, 160, 243, 245
instructions, 119, 140, 219, 261, 269
intellectual property, 187, 279, 307
international donor specifications, 129
internet, 17, 25, 175, 196, 202, 287,
     302, 311
     internet access, 22, 68, 165, 235, 273
inverters, 59, 107–8, 127, 136
investment, 117, 121–2, 139, 195, 197, 215,
     226, 230, 245, 299, 308, 310

Japan, 82

Kenya, 23, 73, 75, 89, 233, 262
kerosene, 9, 16, 18, 96, 167, 203, 274, 310
kiosks, 268, 287
knowledge
     foundation knowledge, 73, 77
     knowledge networks, 268
     knowledge sharing, 53, 73, 81, 117,
          133, 302

labour, division of, 303
landlocked countries, 100, 150, 304
language, 53, 72, 139, 239
LCD (liquid crystal display), 111, 252, 284
LED (light-emitting diode), 40, 76, 88, 96,
     140, 284, 306
leisure, 5, 98, 218, 237
light bulbs, 13, 86–7, 100, 261, 306
lighthouses, 82
Lighting Africa, 96
literacy, 72, 119, 273, 306
loans, 160, 166, 236, 311

local suppliers, 9, 115, 120, 132, 223, 262, 274
logistics, 266, 271, 286, 302, 304
London, 12, 99, 173

mains electricity. *See* grid electricity
maintenance, 6, 64, 85, 87, 100, 122, 134
Malawi, 117, 145, 151, 229, 262
malnutrition, 237
marine applications, 69
market-driven product development, 255
marketing, 42, 79, 95, 102, 121, 136, 139, 156, 191
market sectors, 67
medicine, 159, 212, 282, 308
meters, metering, 49, 52, 86
micro-networks, 268, 272, 275, 287
microprocessors, 113, 154, 206, 209, 248
Middle East, 82
migration, 7, 8, 14, 25, 303–4, 306
military applications, 69
Millennium Development Goals, 163, 308
mini-grids, 25, 65, 85, 106, 134
missionaries, 148, 294
mobile phones
    cell towers, 251
    pay-as-you-go, 86, 223, 263, 270
modules. *See* panels
money, 147, 159, 177, 180, 279, 295, 299
mother of invention, 186
motivation, 19, 152, 155–6, 172, 177, 244, 287
Moyo, Dambisa, 148
multinationals, 185, 197, 200, 202, 214, 238, 250, 267, 282, 293
myth, 234, 258
    of clean fuel, 41
    of low cost, 188
    of purchase cost, 101

NASA, 138
national grids, 12, 13, 66, 88. *See also* central generation
natural disasters, 13, 149, 197
natural resources, 150, 229, 265, 291, 317
Nepal, 151
neutrality, 47, 54, 73, 79, 294
NGO (non-government organisations), 78, 126, 128, 140, 151, 235, 267
Nigeria, 23, 229
non-profit organisations, 139, 147, 152, 190, 223, 224
Northeast Blackout, 6
nuclear, 12, 35

obesity, 176, 189
OEM (Original Equipment Manufacture), 229, 234
offshore platforms, 64, 83, 107

oil and gas, 68, 83, 136, 238, 298
Oman, 62, 65, 99
One Laptop Per Child (OLPC) initiative, 257
operational life, 53, 83, 105–6, 108, 114, 161, 183, 283, 308
opulence, 152, 153–4
orientation (of panels), 60, 66, 83, 270
outages. *See* power cuts
Oxfam, 147, 152

packaging, 75, 79, 119, 284
Papua New Guinea, 160
parking meters, 62, 91, 99
partnerships, 285
    global local networks, 277
    local partner networks, 244, 265
patents. *See* intellectual property
peak demand, 7, 23, 311
people, 73, 233, 258–9, 265, 268, 274, 281
    local people, 244, 257
    off-grid people, 39, 223, 230, 232
performance claims, 66, 91, 105, 119, 140, 199
Personalised Energy, 47, 70, 72, 81, 160, 181, 205, 246, 257, 291, 294
petrol, 33, 308
philanthropy, 98, 149, 155, 223, 306
phones. *See* mobile phones
photoelectric effect, 57
Photovoltaics (PV), 57
    anti-reflective, 82
    arrays, 23, 69, 82, 114–15, 127, 294, 308, 318
    cleaning and maintenance, 50, 73, 78, 85, 162, 249
    crystalline silicon PV, 238
    grid-connected, 58, 66
    history, 82
    legacy of, 87, 92
    off-grid, 66
    off-grid PV systems, 36, 69, 81, 83, 89, 95, 108, 248
    panels, 33, 35, 55, 57, 238, 247, 270
    pico solar, 96
    PV cells, 57, 62, 82, 135–6, 196
pipelines, 64, 83, 298
policy, 20, 86, 163, 294
politics, 13, 37, 147, 225, 291, 293, 296, 312
pollution, 16, 17, 34, 36, 155, 311
population, 145, 147, 172, 221
population density, 18, 100, 150
poverty, 130, 138, 145, 163, 174, 189, 236, 268, 304
    inequality, 152
    numbers of people in, 149, 169
    reasons, 151
power cuts, 6, 21, 250
pre-failure notification, 248, 286, 303

procurement, 125, 200, 224
  specifications, 126
product performance, 98, 140, 250
profit, 59, 63, 197, 210, 212, 265, 287
profit motive, 98, 163, 180, 187, 195,
    198, 277
public confidence, 123
PV panels, 55, 78, 90

quality, 100, 132, 184, 188, 190, 245, 266,
    268, 293, 307
  how to sell, 191
  of consumer electricity, 6, 11, 23
  of stand-alone PV, 118, 133, 160, 199
quality and value, 34, 95, 111, 117, 121,
    162, 243
quality control, 161, 224
quality of products, 39, 63

rate of change, 50
rationing, 52
reactive power, 6
reading light, 258, 261, 278
recycling, 183, 205, 210, 214
redundancy, 5, 26, 84
refrigeration, 67, 101, 208, 278, 280
regional solutions, 64
reliability, 5, 21, 62, 113, 183, 221, 231, 235,
    246, 280
religion, 74, 293
renewables, 17, 60, 164, 183, 229, 234
rental, 269
reputation, 71, 79, 81, 117, 123, 140, 198,
    202, 213, 250, 285
responsibility (personal), 48, 176
risk, 139, 166, 196, 212, 229
rural electrification, 64, 69, 86, 88, 100
Rwanda, xix, 73, 107, 110, 119, 155, 165,
    219, 246, 262

Sachs, Jeffrey, 148, 303
sales, 75, 98, 139, 203, 209, 226, 238,
    266, 272
sanitation, 169, 172, 220, 297
satellites, 61, 68, 82, 293
SCADA, 83
scale, 169
  errors of, 225
  problems of, 173
second-tier consumers, 38
sectors of society, 67
security, 8, 51, 64, 68, 180, 208, 248
self-interest, 178
selfish desire, 179
Selling Daylight, 48, 81, 159, 257, 291,
    293, 295
shelters, temporary or emergency, 64, 69,
    207, 220, 297
shipping, 64, 87, 175, 211, 271, 308

skills (workforce), 24, 28, 86, 221, 261
small island states (SIS), 217, 220, 310, 313
smartphones. *See* mobile phones
social change; social mobility or
    opportunity, 25, 65, 147, 196, 232,
    307, 311
solar home systems (SHS), 64, 65, 97, 112,
    248, 260, 299
solar lanterns, 96, 201, 261
solar thermal, 57, 169
South America, 74, 176, 262
South Korea, 13
space tourism, 295
sports stadiums, 170, 219
stakeholders, 118, 123, 172, 283
stand-alone energy, 54, 55, 61, 70, 81, 90,
    98, 137, 205. *See also* Photovoltaics
    (PV): off-grid PV systems
start-up current, 108
statistics, xx, 148, 151, 229
stem cell research, 295
storyboards, 54, 76, 79, 117, 121
strategy
  commercial, 10, 159, 188, 205, 273, 277,
      301, 313
  long-term, 195, 245, 296, 299
street furniture, 62, 64, 69
street lighting, 88, 126
subsidies, 9, 59, 108, 161, 298
Sundial (product), 75, 233, 259
sun, sunlight, xvii, 76, 111, 138, 211, 251
supply and support networks, 181, 191,
    206, 277
supply chain, 96, 120, 125, 172, 269,
    286, 298
sustainability, 33, 41, 49, 164, 182, 220,
    297–8, 303
SWOT analysis, 201
synchronisation (networks), 235

tariffs (feed-in), 59, 66
technology giants, 185
telecommunications, 61, 64, 68, 82–3, 87,
    117, 128, 286
television, 282
temperature (technical impact of), 29, 115,
    131, 209, 251, 283
tent applications, 69, 207, 218
The local level, 120
torches, 63, 69, 96, 188, 260
tourism, 153, 175
training, 28, 86, 117, 122, 133, 162, 263, 303
tribal loyalties, 149, 265, 273

Uganda, 127, 131, 160
Ugandan Ministry of Education and Sports
    (MoES), 127, 131
unexpected customers, 233
unforeseen challenges, 211

unforeseen circumstance, 273
unforeseen opportunity, 210
United Kingdom, 7, 17, 23, 117, 305
United Nations (UN), 164, 182, 201,
    257, 308
United States of America, 5, 7, 13, 28
unlimited electricity, 32, 42
urbanisation, 25, 29, 75, 154, 268, 280, 305,
    306, 307
USAID, 224
user interface, 49, 77, 110, 116, 155, 165,
    208, 269, 283, 295, 303
user involvement/engagement, 48, 71, 79,
    85, 100, 109, 111
users
    user behaviour, 110, 112
utility grids. *See* grid electricity
utility scale PV, 60

value, 50, 98, 138, 191, 246, 301
    minimum threshold of, 98
    of stand-alone PV, 61, 87, 118, 221
value proposition, 66, 81, 99, 100, 112,
    184, 263
vehicles. *See* cars

versatility, 47, 67, 184, 207, 209, 214, 226,
    230, 271, 310
visual instruction system (VIS), 54, 71, 102,
    111, 121, 198, 206, 283
visual media, 121
vocational learning, 306, 309
voltage (electricity quality), 15, 6, 59, 60,
    131, 206

warranties, 59, 65, 75, 114, 126, 131, 251,
    261, 283
waste, 13
waste (energy), 11, 13, 51, 106
waste (money), 29, 89, 127, 162, 256
water (clean), 171, 173, 175, 219, 309
water pumping, groundwater pumping, 61,
    64, 82, 89
Watts-peak (Wp), 59, 204
weather, extremes of, 61, 88, 117, 247
Wi-Fi, 118, 134, 252, 306, 317
wind power, 23, 33, 35, 231, 267
wood, 9, 12, 15, 18, 163, 170, 231, 309
World Bank, the, 85, 96, 127, 130–31, 139,
    151, 172, 184, 190, 224, 266, 296
World Resources Institute, 201